T0200715

The Voyage of Thought

The Voyage of Thought is a micro-historical and cross-disciplinary analysis of the texts and contexts that informed the remarkable journey of the French ship captain, merchant, and poet, Jean Parmentier, from Dieppe to Sumatra in 1529. In tracing the itinerary of this voyage, Michael Wintroub examines an early attempt by the French to challenge Spanish and Portuguese oceanic hegemony and to carve out an empire in the Indies. He investigates the commercial, cultural, and religious lives of provincial humanists, including their relationship to the classical authorities they revered, the literary culture they cultivated, the techniques of oceanic navigation they pioneered, and the distant peoples with whom they came into contact. Ideal for graduate students and scholars, this journey into the history of science describes the manifold and often contradictory genealogies of the modern in the early modern world.

Michael Wintroub is Associate Professor at the University of California, Berkeley. He authored *A Savage Mirror: Power, Identity and Knowledge in Early Modern France* (2006) and has published widely in journals such as the *American Historical Review*, the *British Journal for the History of Science*, *ISIS*, the *Renaissance Quarterly*, *Annales: Histoire, Sciences Sociale*, and the *Sixteenth Century Journal*. Wintroub has received numerous awards and honours, including grants from the National Endowment for the Humanities, the American Council of Learned Societies, the Mellon Foundation, and the Sixteenth Century Society, where he is a two-time winner of the Nancy Lyman Roelker Prize.

The Voyage of Thought

Navigating Knowledge across the Sixteenth-Century World

Michael Wintroub

University of California, Berkeley

CAMBRIDGE
UNIVERSITY PRESS

University Printing House, Cambridge CB2 8BS, United Kingdom

One Liberty Plaza, 20th Floor, New York, NY 10006, USA

477 Williamstown Road, Port Melbourne, VIC 3207, Australia

4843/24, 2nd Floor, Ansari Road, Daryaganj, Delhi – 110002, India

79 Anson Road, #06-04/06, Singapore 079906

Cambridge University Press is part of the University of Cambridge.

It furthers the University's mission by disseminating knowledge in the pursuit of education, learning, and research at the highest international levels of excellence.

www.cambridge.org
Information on this title: www.cambridge.org/9781107188235
DOI: 10.1017/9781316946459

First published 2017

Printed in the United Kingdom by TJ International Ltd. Padstow Cornwall

A catalogue record for this publication is available from the British Library.

Library of Congress Cataloging-in-Publication Data
Names: Wintroub, Michael, author.
Title: The voyage of thought : navigating knowledge across the sixteenth-century world / Michael Wintroub.
Description: Cambridge, United Kingdom ; New York, NY : Cambridge University Press, 2017. |
Includes bibliographical references and index.
Identifiers: LCCN 2017008195 | ISBN 9781107188235 (hardback)
Subjects: LCSH: Parmentier, Jean, –1529. | Voyages and travels–History–16th century. | Scientific expeditions–History–16th century. | Discoveries in geography–French. | Explorers–France–Biography. |
BISAC: TECHNOLOGY & ENGINEERING / History.
Classification: LCC G440.P216 W56 2017 | DDC 910.4/5–dc23
LC record available at https://lccn.loc.gov/2017008195

ISBN 978-1-107-18823-5 Hardback

Contents

Figures

Every effort has been made to contact the relevant copyright-holders for
the images reproduced in this book. In the event of any error, the publisher
will be pleased to make corrections in any reprints or future editions.

Acknowledgments

I would like to thank a number of colleagues, friends and students for having read and commented on various chapters of this book; in particular, Hélène Mialet (without whom I could never have imagined, let alone written, this book), Beate Fricke, Mario Wimmer, Jessie Hock, Jeffrey Hadler, Simon Schaffer, David White, John Ødemark, Guillaume and Isabelle Sannie, Tony Sandset, Ron Makleff, Anooj Kansara, Alexander Arroyo and Gloria Yu. I would also like to thank the Townsend Center at UC Berkeley whose fellowship support enabled me to dig into the writing of several of the book's chapters. A grant from the National Endowment for the Humanities allowed me to complete the manuscript. Chapters 2 and 3 have been published in somewhat different form, in the *British Journal for the History of Science* and the *Renaissance Quarterly*, respectively.

Introduction

We are on the confines of the Frozen Sea, on which, about the beginning of last winter, happened a great and bloody fight between the
Arimaspians and the Nephelibates. Then the words and cries of men
and women, the hacking, slashing, and hewing of battle-axes, the shocking, knocking, and jolting of armours and harnesses, the neighing of
horses, and all other martial din and noise, froze in the air; and now, the
rigour of the winter being over, by the succeeding serenity and warmth
of the weather they melt and are heard.

– François Rabelais[1]

Out of sight of land for days, weeks and sometimes months, intrepid men
on leaky ships sailed dangerous waters in the hunt for pepper, ginger,
gold and souls. Along the way, they collected intelligence about themselves, about others, and about the world: where to find water, food,
safe ports-of-call, valuable commodities, dangerous reefs, good winds,
strong currents, and the height of the stars. New techniques were tried
and instruments used; information was collected and refined into rituals
of standard practice and inscribed onto paper, wood, and metal; onto
the bodies of experienced sailors; and into the design of the ships that
carried them. This routinization was associated with the establishment
of trade routes over impossibly long distances, and eventually into disciplines seemingly as far removed from standards of navigational practice
as archaeology, history, geography and anthropology. This book is about
such journeys. It is a charting of the physical trajectory of two ships that
sailed from Northern France to the Indonesian island of Sumatra and
back again early in the sixteenth century. It is also about the charting of
attempts to operationalize metaphors, appropriate space, and discipline
men and their relations with the world through an examination of their
modes of action, their styles of thought, and their manifold entanglements with techné and technology, nature and things. These two ships,

[1] François Rabelais, *Five books of the lives, heroic deeds and sayings of Gargantua and his
son Pantagruel*, trans. Sir Thomas Urquhart of Cromarty and Peter Antony Motteux
(New York, 2005), 563.

the relatively small and more maneuverable Le Sacre, and the larger, La Pensée, were sailed across the seas by Jean Parmentier and his younger brother, Raoul. Their expedition was sponsored by the maritime kingpin, Jean Ango, an owner of ships and employer of sailors, a banker and financial backer – an *armateur* and entrepreneur – who put together consortiums to finance and provision long-distance voyages up and down the Atlantic coast; from Flanders to the North, and Africa to the South, and across the seas to France Antarctique (Brazil), Terre-Neuve (Canada), Guinée, the Antilles and the Indes orientales.[2]

Dieppe, by the standards of the day, was a relatively large city, having a population of approximately 15,000 in 1550.[3] Its size was matched by the importance of its port. Ango harnessed his status in Dieppe and beyond to the success of his maritime ventures, whether in cloth and spice in Antwerp, brazilwood in Brazil, cod in Acadia or booty captured from Portuguese, Spanish, Flemish and English ships by his fleet of pirate-privateers.[4] With his profits, he bought land and titles (the domaine de Varengeville and the fiefs de la Rivière à Offranville, Desmaillets, Saint-Pierre l'Advis, Sainte-Marguerite, and Gerponville), as well as offices (Grènetier-Receveur de la Vicomté, Conseiller de la Ville, Vicomte, Capitaine du château pour le Roi, and Gouverneur de Dieppe).[5] At the height of his career he was in charge of virtually every aspect of Dieppe's political and fiscal governance. Along with money, he acquired taste, sponsoring artists, mapmakers, poets, sculptors, and translators. His power flitted between his fleet of ships and his cultural sophistication to grow into a reputation that merited the friendship of the king's sister, Marguerite de Navarre, and even a visit by the king himself – who he entertained in his spectacular quayside home, named, like his ship, La Pensée.[6]

Embellished with sculpture and bas-reliefs carved in wood and plated with gold, and decorated with furniture, sculpture and paintings of the

---

[2] On Ango see Paul Gaffarel, *Jean Ango* (Rouen, 1889); Gabriel Gravier, *Jean Ango: Vicomte de Dieppe* (Rouen, 1903); and Eugène Guénin, *Ango et ses pilotes* (Paris, 1901). Indes, here, will refer to the *Indes orientales*; Indies, on the other hand, will refer to the lands discovered to the west.

[3] Philip Benedict, "French Cities from the Sixteenth Century to the Revolution: An overview," in Philip Benedict (ed.), *Cities and Social Change in Early Modern France* (London and New York, 1992), 8.

[4] Ango had at least 60 ships; as many as 30 more were mentioned, in passing, in Portuguese archives. See Michel Mollat, *Le commerce maritime normand á la fin du Moyen Age: Étude d'histoire économique et sociale* (Paris, 1952), 501. On Ango's sponsorship of acts of piracy see ibid.

[5] Ibid., 506.

[6] See F. Génin (ed.), *Lettres de Marguerite d'Angoulême, soeur de François Ier, Reine de Navare* (Paris, 1841), 252–255.

very best quality, La Pensée was reputed to be the finest wooden home in all of France.[7] It was also the hub of a wide-ranging merchant empire, a clearing-house for navigational information, an outpost for humanist learning in the provinces, and a meeting place for mapmakers, sailors, poets, pirates, and privateers. Through its namesake (and other ships like it), La Pensée was connected to the world. It was, quite literally, "a thought" extended along networks of maritime trade and exploration that spanned the globe. Thus, though La Pensée might have referred to a violet or yellow flower – a "pansy" – it most certainly embraced its homonym, a "thought."[8] Indeed, between flowers and thought there was less daylight than one might think, as pansies were emblematic of thoughts, dreams, and memories.[9] As Ophelia, in Shakespeare's *Hamlet*, said: "And there is pansies, that's for thoughts."[10] Derived from the Latin, *pendere* or *pensare*, pensée thus referred not only to a flower, but to think, to reflect, to be poised and deliberate, to judge, to consider carefully, and to weigh. It is interesting to note in this regard that the meaning of the English verb "to weigh" – "to bear from one place to another; to carry, to transport" – has an uncanny similarity to the meaning of translation, which derives from *trans*, meaning to cross over or go beyond, and *fero*, meaning to bear or carry.[11] One can thus imagine La Pensée, as both home and ship together, forming a material-neural network of translation along which Ango could take the measure of the world and his progress within it. Put another way, it was through the combined (inter) actions and displacements – the ongoing, and back and forth, translations – of these "thoughts" that Ango was able to leverage and mediate enormous geographic and social distance so as to materialize his dreams for profit, glory and respectability, whether in ginger, pepper and gold, or the visit of kings.

In 1529 two of Ango's men, the brothers Parmentier, led a mission to the Indes in search of profit and glory; it was a third of a trilogy of failed voyages to find spice in the Far East, the first two having never returned. They set off on Easter Day and sailed for seven months before reaching what would be their final destination, the village of Ticou on

[7] Guénin, *Ango et ses pilotes*, 7.
[8] For Randle Cotgrave, "la pensée" was "a thought, supposal, conjecture, surmise, cogitation, imagination; one's heart, mind, inward conceit, opinion, fancie, or judgment; also, the flower Paunsie." See Cotgrave, *A Dictionarie of the French and English Tongues* (1611), s.v. *pensée*.
[9] See Pamela Porter, *Courtly Love in Medieval Manuscripts* (Toronto, 2003), 11.
[10] William Shakespeare, *The Tragicall Historie of Hamlet Prince of Denmarke* (London, 1603), Act 4 Scene 5, fol. H^v.
[11] *Oxford English Dictionary*, 2nd edn (Oxford, 1989), also available as the OED Online, s.v. translation.

the central west coast of Sumatra. Little trace remains of their voyage. There is a shipboard journal, and there is poetry.[12] Jean Parmentier, and his *cosmographe* (his navigator), Pierre Crignon, were, as it turns out, not only famous merchant-explorers who led ships to Brazil, Guinée, La Terre-Neuve and the Indes orientales, they were also poets of renown.[13] More than just poets, however, they were men of learning steeped in the classics. In 1528 Parmentier translated *The Catiline Conspiracy* by the Roman historian, Sallust, for his patron, Jean Ango; he was working on a translation of the *War with Jugurtha* for François I when he died on board La Pensée in 1530. The poems they wrote were crowded with literary allusions to Roman and Greek history, philosophy, and mythology, with references to Pliny, Pythagoras and Cicero rubbing shoulders with Phebus, Pallas and Neptune. Just as importantly, their poetry vividly recalls their lives at sea. They wrote of navigational instruments and maps; sailing techniques and astronomical observations; the power of the oceans, the contingencies of the weather and the constancy of the stars. Their words gave voice to lives entangled in ropes, sails, and ships harnessing the winds; at the same time, their poems sang of a longing for God's guidance and of their devotion to the Virgin Mary, mother of God. In intermingling these social, spiritual and natural worlds, their verse translated ships into vessels of "sovereign beauty" that would transport humanity "to the sacred port where glory abounds";[14] into navigational instruments, like astrolabes, that mirrored the "perfect symmetry of the Virgin" as drawn by God's "error free compass";[15] and into world maps and the siting of the pole star that would guide wayward and desperate sailors to both God and to profit.[16] In their poetic prayers and theological verse one can also discern sophisticated and recondite mathematical techniques; knowledge of geography and geodesy, and the practical skills and bodily techniques of expert sailors. We can additionally find in their obscure and difficult to read verse, displays of linguistic mastery over the complex and esoteric grammatical rules that governed the writing of poetry in the fashion of the *Rhétoriqueurs*.[17] Their poems were read out

---

[12] See John Nothnagle, *Pierre Crignon: Poète et navigateur. Oeuvres en prose et en vers* (Birmingham, AL, 1990), hereafter, Crignon; and F. Ferrand (ed.), *Jean Parmentier, Oeuvres poétiques* (Geneva, 1971), hereafter, Parmentier.

[13] See Chapter 2.

[14] Parmentier, *Oeuvres*, 27–29.

[15] Ibid., 62–65.

[16] Ibid., 25.

[17] See, for example, Pierre Fabri's *Le Grant et vray art de pleine rethorique* (Rouen, 1534). See Chapter 2, and Michael Wintroub, *A Savage Mirror: Power, Identity and Knowledge in Early Modern France* (Stanford, 2006), esp. chapter 4. More generally, see Paul Zumthor, *Le masque et la lumière: la poètique des rhètoriqueurs* (Paris, 1978).

to audiences and judges of like-minded merchants, sailors, humanists, priests and nobles in several different Norman poetry confraternities; and, as we will see, one was even read aboard La Pensée as a means of inspiring her hard-pressed crew to continue on their intrepid course across dangerous and unknown waters. For all these reasons, the poetry of Crignon and Parmentier will help orient us by mapping crucial stages of La Pensée's and Le Sacre's voyage; it will also provide clues to the more expansive – social, epistemic and spiritual – nature of their journey.

The Voyage of Thought is thus not simply a story about the relatively little-known early attempts by the French to challenge Portuguese oceanic hegemony and carve out a trading empire in the Indes; it is also an investigation into the commercial, cultural and religious lives of provincial humanists in Dieppe, examining their relationship to the classical authorities they revered, the literary culture they cultivated, the techniques of oceanic navigation they pioneered, the distant peoples that they met, and the ways in which all these different ideas, practices and values were wrapped up in histories of spiritual and rhetorical practice as evinced by the poetry they wrote and publically recited on the feast days of the Immaculate Conception and the Assumption.

Each chapter of this book is based on a moment of La Pensée's and Le Sacre's journey, from its inception to its return; and each is oriented around a word: *Information, Expertise, Translation, Scale, Confidence* and *Replication*. My aim in pursuing this itinerary is not simply to explore the physical trajectory of these Norman merchants as they crossed the world in search of profit, glory and redemption, but to trace the specifically early modern resonance of ideas and practices that have come to be so important to our present-day understandings of science and its history. Far from being an attempt to locate modern practices and modes of thought and action in the early modern world however – my aim here will be to explore the particular historical, social, and cultural etymologies of these "key" words at a time when their meanings were far from fixed and settled.[18]

---

[18] Put another way, I propose we pursue a kind of "connected history" by following a series of words – Information, Expertise, Translation, Scale, Confidence and Replication – down the interweaving rabbit holes of their diverse and related meanings. We will thus trace entanglements of religion, politics, commerce, culture, class and classification schemes as they intersected, broke away from and/or reinforced each other in practice – that is, on ships crossing oceans, led by poets armed with the latest methods, to trade (not so successfully) with peoples living on the other side of the world. See Sanjay Subrahmanyam, "Connected Histories: Notes towards a Reconfiguration of Early Modern Eurasia," in Victor Lieberman (ed.), *Beyond Binary Histories: Re-imagining Eurasia to c. 1830*, (Ann Arbor, 1999), 289–316; Sanjay Subrahmanyam, *Explorations in Connected History: From the Tagus to the Ganges* (Oxford, 2005); Kapil Raj, *Relocating Modern Science: Circulation and the Construction of Knowledge in South Asia and Europe, 1650–1900* (Basingstoke and

"No man is an *Iland*, intire of it selfe; every man is a peece of the *Continent*, a part of the *maine*," said John Donne in the early seventeenth century; one could say the same for a word. Meaning, as enacted in the world, is ambiguous and agonistic; it is situational, though mobile (with a great deal of work); it is distributed, fractured, argued over and inflected; it is caught in webs tremulously vibrating across space and time, between people and peoples, and it is entangled with things, techniques, interests, bodies in action, affect, and, of course, with other words. Rabelais knew this well when he described the sound of frozen words discovered on glacial seas thawing in the spring: a cacophony of sounds – the sounds of a battle between different voices: men, women, horses and the clashing spears – all muddled together into unintelligible gibberish. The idea of fixing such words, of encasing them between the covers of definitive collections as artifacts frozen outside of time, place or interest, might have been the dictionary's promise and the lexicographer's dream, but for a man like Rabelais it would have been, literally, nonsense.[19]

Rabelais was perched at the edge of a divide in the making – a divide composed by dreams of order, purity and truth wending themselves into a world that was discontinuous, halting, filled with contradictions and reversals, mercurial desires and unpredictable fears. His contemporaries, Jean Ango, Jean Parmentier, and Pierre Crignon, were similarly positioned, though with a different point of view. These were ambitious, anxious and apprehensive men who sought to collect and give order to their experiences in the world – whether it be of language, customs, manners and dress, the directions of the winds, the geography of the human body, or the height of the stars. Theirs was a form of life based on mastering distance and translations of scale: the idea that the impossibly large, the unimaginably small, the abstract, the contingent, the particular, and the invisible, might be translated – contained, constrained, and disciplined – into models, instruments, practices, representations and inscriptions that could be gauged and engaged – "up close and faraway" – by observant eyes and practiced hands.

New forms of civility and cultural status, and new forms of territorial and administrative power, were implicated in these acts of observation and inscription, of setting out, setting down, and making mobile. Dictionaries and grammars, for example, were, like maps, a projection

---

New York, 2007); Amit Prasad, *Imperial Technoscience: Transnational Histories of the MRI in the United States, Britain, and India* (Cambridge, MA, 2014), esp. 1–9; and Michael Wintroub, "Translations: Words, Things, Going-Native and Staying True," *American Historical Review* 120:4 (October 2015): 1185–1217.

[19] And thus, surely, a commentary on his own practices as an author.

of distance; they were a distanciation and a decontextualization; they were ordered geographies of languages as spoken and written and translated into metrologies of appropriate linguistic usage (and social prestige).[20] Yet alongside dictionary definitions and grammatical rules, usage resisted; as Rabelais' description of frozen words illustrates, words were voiced in unredeemable contexts, in elusive interactions, in inflected narratives, in juxtapositions and combinations with other words, with things, peoples and situations that made their meanings ambiguous, paradoxical and contradictory. Thus, while this book will focus on attempts to stabilize and discipline contingent and singular experiences into definitions, algorithms, recipes, standard practices, and maps capable of transforming voyages of geographical discovery into regularized trade routes, and voyages across epistemic space into definitive expressions of social distance and authority, it will also be attentive to the precariousness of such translations. Rabelais was right, the meaning of words as used, spoken, written about and understood was ineluctably contextual, trapped in complex interactions, and entangled in environments of interacting peoples, animals, things and interests: a battle indeed, but one lost in time and in space. The aim of the present work will be to reanimate some of these early modern battles among men, technologies and nature; to give them back their contingency, their aspirations, their horror, their joy, and their pain; and by so doing to chart a kind of historical topology of meanings made in practice: a *Begriffsgeschichte* of the purifying impulse of the modern.[21]

A generation before Rabelais wrote about Pantagruel's frozen words, Jean Parmentier and his brother Raoul set off for the other side of the world; they had only primitive and inaccurate maps; their translations were never sure and were often wrong; their movements across scale struggled to find sure measure; confidence was difficult to come by, and trust was hard to win; at the end of each day, they had a hope and a prayer, but they had no idea if they would succeed in reaching their final destination, or return home again. This book is about how such hopes and prayers were historically operationalized in the coordination of men, ships, seas and stars such that one voyage across the world could become many, and many worlds could become one.

---

[20] In this sense, we can perhaps view metrology as a kind of naturalization of social hierarchy.

[21] See especially, Bruno Latour, *We Have Never Been Modern* (Cambridge, MA, 1993).

1 Information: Pilgrimage in a Church of Poems

We live in the "Information Age." Few would argue with this, but the word, "information," connotes different things to different people. In France, for example, it might mean the nightly news, or more generally the *presse d'information* (the news media). In an ideal world, the news would be an unbiased recounting of the "facts" of contemporaneous events; however, even the most biased adamantly claim to live in an "objective" world. Recall, for example, that France's *Ministère de l'Information* from 1942–4 was the Nazi collaborator, Pierre Laval. It is thus perhaps not surprising that information is so often defined today in terms of its opposition to knowledge and understanding. This can be linked not only to its association with propaganda and bias (and, of course, to men like Laval), but also with putative attempts to escape from them. Information here is taken in the sense of "raw" and "neutral" data rather than opinion, interpretation, knowledge or analysis. This has not only had the effect of producing a supposed distance between information and bias, but also of establishing a disjuncture between information and knowledge. Put another way, the meaning of "information" has flown into and merged with the data in which we are all (presumably) drowning. This opposition, as a possibility, derives from a striking reversal in the historical meaning of the word. The reification of data – as facts deracinated from the schema, the metaphors, and the knowledge-interests that created their conditions of possibility – has, in this sense, bled into the concept itself, disrupting, or rather, weighting our understanding of information toward one particular side of its multifaceted meaning. Indeed, if we track information's roots back in time, the tensions inhabiting its current meanings become both more pronounced and more complicated.

From the Latin *informatio*, information was the formation, creation, teaching, or arrangement of knowledge, as in the "instruction" or "formation" of a supplicant, a student, a citizen, or a criminal.[1] More abstractly, it was also understood as that which gives an idea or a concept

---

[1] OED, s.v. "Information."

8

form. Among the earliest meanings of information was juridical: to conduct a criminal inquest – *"faire des enformacions."*[2] The collection of information, in this sense, was part of a process of policing. An inquest was
both a retrieval of knowledge (to be informed about), and an imparting
of knowledge – to (re)form or instruct. Put somewhat differently, information implies a hermeneutics of control: to know so as to inculcate,
instill, reform, subdue. In this sense, information flows simultaneously in
two directions: from the weak (e.g. criminals and heretics) to the powerful (e.g. inquisitors and jurists), and then back from the powerful to
the weak. The reputations of the subjects of inquisition necessarily colored information's meaning, as did the less than exemplary reputations
of those associated with the gathering and communication of this intelligence – the informants, the eyewitnesses, the spies, the snitches, the
stoolpigeons and the traitors, and the others (the many others) who were
compelled to give up their information, whether by persuasion or force.
Randle Cotgrave's *A Dictionarie of the French and English Tongues* (1611)
documents some of these pejorative associations: "informe," he tells us,
is "shapeless, ill-favoured, fashionless; ouglie, rude."[3]

Information was not only human, it also had material instantiations
as "evidence" – that is as "data." On the one hand, such evidence could
testify when there were no human witnesses available; on the other, evidence could serve to replace and/or discipline the testimony of humans
who were prone to exaggerate, selectively edit, misrepresent or lie. Proof
was here embodied in discrete items that could be seen, touched, examined, collected and moved; thus would a judge be instructed by the
perusal of incriminating letters, maps, books, and perhaps even dead
bodies to form an argument of guilt or innocence.

The credibility of information, as either proof or deception (*mésin-
formation*), in its most stark formulation, tracked closely to traditional
notions of social place and power; in juridical and theological contexts, it was also inseparable from a great deal of violence and coercion.
A source's trustworthiness was, in this sense, a condition of perceived
social reliability: priests and nobles were more credible than merchants,
Jews, soldiers, sailors, heretics, and spies.[4] But means could also be
employed to ensure the veracity of information obtained from less creditable sources: blackmail, threats and coercion (e.g. judicial torture).

[2] *Trésor de la langue française*, s.v. *information*.
[3] Cotgrave, *A Dictionarie*, s.v. *information*.
[4] See, for example, Lisa Voigt, *Writing Captivity in the Early Modern Atlantic: Circulations
of Knowledge and Authority in the Iberian and English Imperial Worlds* (Chapel Hill, 2009),
41; and Andrea Frisch, *The Invention of the Eyewitness: Witnessing and Testimony in Early
Modern France* (Chapel Hill, 2004), for example, chapter 2.

Information was necessarily intertwined with power. If it was to be reliable, information and its sources needed to be closely monitored, checked and disciplined.

The judicial hunt for information against murderers and thieves was, as we have noted, called an inquest, which in turn, derives from the Latin, *quaesitus* or *quaero*, to seek, to inquire, to hunt. From the twelfth century, it came to be closely associated with inquisition – the interrogation – of faith. Indeed, information could derive from supernatural as well as human sources; most notably, the divine inspiration of the Holy Spirit.[5] As Lydgate put it, Christ was "First a prophete by holy informacion."[6] But just as human informants were not always trustworthy, so too supernatural sources needed to be carefully vetted. Indeed, how was one to be sure that such sources of instruction did not derive from the devil himself? Information, as inquisition, was the process whereby unbelievers, heretics and criminals could be identified by the testimony of informers; it was then employed to inform, instruct and exemplarily prosecute those who had been denounced.[7] Information was thus closely allied with governmentality through the hunt for wayward and "ouglie" souls, miscreants, heretics, and erstwhile subjects that were to be fashioned, (in)formed, disciplined, and instructed.

Inquisition (*enquête*) received a slightly different inflection in accounts of the chivalrous exploits – the *quests* – undertaken by Arthurian knights. These quests, like the persecutory inquisitions of heretics, aimed at proving faith by the exercise of heroic deeds – e.g. the search for perfect (chaste) love, the pursuit of the Holy Grail, or the defense of the faith against Saracen invaders. A quest, however, could also take more worldly forms, namely, the search for objects of great symbolic and/or material value. Such quests were frequently infused with religious significance, as for example the medieval French cleric and trader (thief) of relics, known to us today only as Felix who plied his trade less as a form of religious devotion than for profit, that is, as a "*questus causa*."[8] A "quest for profit" could thus be inflected in a spiritual direction, lending legitimacy and status to more worldly forms of trade. Jean Parmentier, for example, wrote (c. 1527) of a ship called La Marie that voyaged to faraway countries to bring back a "beautiful belly full of rich red wood … for the great profit of all humanity."[9]

[5] OED, s.v. "Information."
[6] Ibid.
[7] R. I. Moore, *The War on Heresy* (Cambridge, MA, 2012), 206.
[8] Michael McCormick, *Origins of the European Economy: Communications and Commerce AD 300–900, Parts 300–900* (Cambridge, 2001), 284, n. 4.
[9] Parmentier, *Oeuvres*, 27–29.

Énquête has its roots in the hunt, and in particular, the barking of dogs at the sight of prey; as the Norman poet Gace de La Buigne put it: "*chercher la bête ou les traces de bête avec les chiens courants.*"[10] It is interesting to note, in this regard, that though clerics were forbidden to hunt,[11] those most closely associated with the founding of the Inquisition, the Dominican order, were known – after their name – as the "Dogs of the Lord" (*Domini canes*). While Dominicans could only hunt for the heresy hiding in the hearts of men, nobles saw hunting as a manly quest for symbolic capital; it was viewed as the exclusive purview of their class. As Stuart Carroll has put it, hunting was the "cornerstone of noble sociability."[12] Hunting for food by the lower classes, on the other hand was, quite simply, poaching. Like the clerical inquisitions of heretics, there was a great deal of discipline and setting apart in the noble hunt, which sought to monopolize hunting by the establishment of reserves, parks and warrens, and by the enactment of laws that strictly forbade – with consequences for life or limb – hunting by non-nobles.

Information, far from simply signifying the datum, *les données*, of the material world as it often does today, was directed at searching them out, acquiring and subduing them. In these senses, it was closely related to *enquête*'s cognate: *querir* – to look, to search for, to seek. It was also closely related to *requérir* (to seek, search, look after, hunt for, request, beseech, implore, or to demand, as in the *Requerimiento* (1513) read by Spanish conquistadors to uncomprehending natives throughout the New World); *acquérir* (to acquire, find, obtain, purchase, to search for, to claim); and, of course, *conquérir* (to conquer, possess, vanquish, overcome, but also to get, to purchase, to win or gain).[13] Material, spiritual and political meanings thus intermingled among the roots of information. While on the one hand it was the act of hunting for and seeking out, it also moved to embrace its prey (its objects): profit, truth and power as well as their expression in commerce, religious practice and colonial acts of domination (whether against law-breakers and heretics at home, or savages in New and faraway worlds). There was, and perhaps still is, something deeply imperial about information.

The material, quantifiable nature of information – as in "a piece" of information, data or knowledge – can be identified as early as the fourteenth century, but it reaches fuller definition by the nineteenth, and

---

[10] Gace de La Buigne, *Roman des Deduis*, 2890 ds T.-L., as quoted in the *Trésor de la langue française*, s.v. *quérir*. See also the OED, s.v. quest.

[11] See, for example, The Fourth Lateran Council, canon XV: "We interdict hunting or hawking to all clerics."

[12] Stuart Carroll, *Blood and Violence in Early Modern France* (Oxford, 2006), 62ff.

[13] Cotgrave, *A Dictionarie*, s.v. *conquérir*.

finds critical mass in the late twentieth, i.e. in our so-called "Age of Information." One can see traces of this meaning, however – as "raw" data – in definitions provided by early modern dictionaries. Information, in this context, refers not only to informing, instructing or investigating but, as Cotgrave intimates, something "shapeless" or "rude," or as Estienne said in his *Dictionarium Latinogallicum* (1552), citing Cicero, as the unsullied, the primitive, the raw form of intelligence: it is the "savage (*rude*) imagination which isn't at all fashioned" (*Imagination rude, qui n'est pas du tout faconnee*).[14] The solidification – the materialization – of information as raw, unfashioned, rude, savage, data that could be decontextualized as discrete "bits," was surely tied to the importance of information, and of record-keeping, for matters of commerce, law and administration. Perhaps it was also related to the concern that the words of informants, and indeed, the textual artifacts of antiquity, might be less trustworthy than the brute reality of things – e.g. coins, medals, seals, as well as kidnapped natives, artifacts, postmortem examinations and exotic commodities.[15]

Information about distant places continued to dance between the persuasive powers of the marvelous and the practical judgments and calculating observations of merchants, explorers, and sailors. Flowing alongside the exoticism of wonder cabinets, travellers' tales and *mappae mundi*, were collections of seagoing and proto-ethnographic information to be found in logs, rutters and charts, and also, of course, in the embodied experience of navigators, pilots, merchants, and sailors who frequented ports up and down the Atlantic coast. Themes of conquest and plunder in exotic lands were characteristic of these quests, whether as chivalrous romances, travel accounts, or cabinets of curiosity.

Just as wonders, monsters and prodigies were often associated with the credulity of the lower orders, so too, the collection of empirical data, however useful, was associated with physical labor, the fleeting material world of particulars, and with those who stood at the bottom of the social hierarchy. The meaning of information straddled these tensions, tipping this way and that, from the collection and study of empirical phenomena and wonderful singularities as data (as intelligence), to information as an organizing principle (*scientia*) that could give form to the formless. Information was both wax and seal. The social tensions implicit in this relation were mediated first by wonders and curiosities, and then, with growing success, by the conversion of information into re-contextualized

---

[14] Robert Estienne, *Dictionarium latinogallicum*, s.v. *Informatio.*
[15] A. Momigliano, "Ancient history and the antiquarian," *Journal of the Warburg and Courtauld Institutes* 13 (1950): 285–315.

raw material – data – to be collected and arranged in treasuries, archives, libraries, and cabinets, and then later in museums, academies and trading houses. Beyond the social-cognitive tensions encapsulated by these various definitions of information, what is of particular interest is the ways in which they were all mediated as expressions of a colonial-mercantile-missionary-administrative ideology; in one way or another, they were all implicated in a hunt – a quest, for land, goods, and souls.

In the pages that follow, we will begin by examining information collected and carried in print that may have informed Parmentier's journey to Sumatra. Information's reach, however, was not simply in intelligence gathered and inscribed, it also embraced processes of mobilization and deployment. We will thus follow the informational nexus of hunt, quest and conquest through its circulation in networked assemblages beginning with, and extending from, Ango's quayside home, called La Pensée (like Parmentier's ship), into Dieppe's most important place of worship, l'église Saint-Jacques, and on to its otherworldly patron, Saint-Jacques le Majeur. We will then follow these extended informational networks into the mnemonic/associational devices employed in the church to make spiritual, meditative, and geographical voyages possible, tracing them both to their uses by humanist circles associated with Ango's business interests in Italy, and to widely disseminated and popular Northern European texts, such as Erasmus' *Colloquies*. We will then pursue a series of inscriptions originating in the Church – in its sculptural program, in graffiti, and in poetry – as a means of fleshing out the materiality of information's entanglement with long-distance exploration and trade.

Spies, Maps and Facts

Expeditions made for profit, adventure and God, such as the one launched by Jean Parmentier and his brother in 1529, were also quests for information, as demonstrated by Pierre Crignon's Log of the voyage, which carefully recorded information about winds, latitudes, faraway lands, and faraway peoples. Crignon's text, however, was not simply a day-by-day recounting of the expedition's progress toward Sumatra; it was an anticipatory act of preparation for the next voyage. The routinization of the inaugural event of discovery into repeatable voyages depended on records supplied by voyages that had already been completed.[16] It also depended

---

[16] See Bruno Latour, *Science in Action: How to follow scientists and engineers through society* (Cambridge, MA., 1987), 215–257, and his essay, "Visualization and Cognition," *Knowledge and Society* 6 (1986): 1–40.

on reasons to "go again," that is, on desires to be satisfied, expectations to be fulfilled, and goals to be achieved. For example, Jean Parmentier stated in his "Oration on the marvels of God and the nobility of man" (a poem he wrote and read aloud to his men while on board La Pensée in 1529), that his voyage was made for God, king and country;[17] yet despite his poetic gloss, as we will see, it was also made for gold, spice, converts, and glory. Whatever the case might be, prior experience – whether the embodied presence of informants who had already been there, or books, maps, instruments, hearsay and rumor – was the *sine qua non* of overseas exploration and trade. As might be expected, the port of Dieppe was a crossroads of such information.[18] Spanish, Portuguese, Flemish, English and Italian sailors could be found there; some were perhaps spies, others were hired hands, others were taken as captives by Jean Ango's predatory pilots; indeed, experience was an invaluable commodity, well worth acquiring whether by money or violence if the opportunity presented itself.[19] Intelligence gathered in Dieppe by Portuguese spies and sent back to Lisbon provides both a sense of how information traveled and the urgency motivating its collection. Thus dispatches from Portugal's ambassador in France, João da Silveira, recount the exploits of one of Ango's most audacious pilots, Jean Fleury, who set off from Dieppe in 1524 with eight ships: L'Espagnole, le Pitipança, la Salamandre, la Citare, la Lingoteira, la Marie, le Papa et le Dragon.[20] Remarkably, Silveira's intelligence also included the names of Ango's partners in backing Fleury's pirate fleet: Nicolas Morel, Guyon d'Etimauville, Belleville [Cardin d'Esquiville-Bléville?], Michel Feré, Silvestre Billes, and several associates in Tours and La Rochelle.[21] The quality of the information collected by Portuguese spies in Dieppe strongly suggests that they had someone "on the inside" who could provide detailed intelligence about Ango's activities; it is also indicative of how seriously the Portuguese took

---

[17] Parmentier, *Oeuvres*, 92.
[18] The "international" array of coins in use in early sixteenth-century Normandy was surely indicative of this; see Chapter 5. Crignon's Log refers to crew members of La Pensée and Le Sacre as Basque, Flemish, Scottish, and Portuguese. On this point see Pablo E. Pérez-Mallaína, *Spain's Men of the Sea: Daily Life on the Indies Fleets in the Sixteenth Century*, translated by C. R. Philips (Baltimore, 1998).
[19] Francis Drake would have fitted in well with Ango's pilots. As Lisa Voigt documents, Drake had an insatiable desire for information acquired through piracy – that is, for the booty of maps, charts, papers, dispatches, and instruments (e.g. Nuno da Silva's astrolabe). A summary of Silva's deposition before the Inquisition (for heresies committed while Drake's captive) indicates that Drake carefully checked, and modified where necessary, this purloined information against his own experience. Voigt, *Writing Captivity*, 256.
[20] See Mollat, *Le commerce*, 501.
[21] Ibid., 501–502.

the threat posed by Normandy's merchant-pirate-explorers.[22] Fleury had succeeded in capturing some thirty Spanish and Portuguese ships that year alone.[23] A January 16, 1530 letter from the João III to Silveira makes specific mention of French corsaires having taken more than 300 Portuguese ships. Of particular interest, in the present context, was the capture of two Portuguese carracks returning from the Indes charged with spice, silk, gold and precious stones that the Portuguese valued at more than 400,000 ducats.[24] Surely, these ships also held instruments, charts, sailors, pilots, go-betweens and translators that could provide valuable information about the lands they had just visited.

The reach of Ango's men extended up and down the Atlantic coast as far as the New World; not only did they attack the Portuguese, but also Spanish, English, and Flemish vessels. One of their most spectacular exploits occurred in 1522 when Fleury succeeded in carrying away two of the three ships transporting the spoils of Cortés' conquest of Mexico back to Emperor Charles V. The treasure that Fleury and his men confronted in the holds of the Spanish vessels must have been shocking. Though news of Cortés' exploits had arrived in Spain as early as 1519, the enormity of Aztec wealth was largely unknown. In addition to gold, the captured booty included emeralds, pearls, the jewels of Moctezuma, the bone of a giant, and three jaguars.[25] Fleury's spectacular and spectacularly profitable act of piracy was commemorated by members of his crew with the *ex voto* offering of stained glass windows at *l'église Saint Martin de Villequier* in 1523. In

[22] Regarding acts of piracy by Ango and his men see, for example, John W. Blake, *West Africa: Quest for God and Gold* (London, 1977), 106, and Édouard Gosselin, *Documents authentiques et inédits pour servir à l'histoire de la Marine normande et du commerce Rouennais pendant les XVIe et XVIIe siècles* (Rouen, 1876), 157–158. Of course, the distinctions between privateers, pirates, and merchants backed by the authority of the French king were ambiguous. In general, a privateer had the institutional sanction of a letter of marque and was not responsible for his actions; rather, he was an agent of the state in an official act of war or reprisal. Piracy, on the other hand, was an individual and indiscriminate act with no official sanction. Having said this, what for a Frenchman was a legitimate competition for markets, goods, and the freedom of the seas, was for the Spanish and Portuguese simply piracy. See Michel Mollat, *Études d'histoire maritime: 1938–75* (Torino, 1977), 473–486; Janice Thompson, *Mercenaries, Pirates, and Sovereigns: State-Building and Extraterritorial Violence in Early Modern Europe* (Princeton, 1996), 22; Michael Kempe, "'Even in the Remotest Corners of the World': Globalized Piracy and International Law, 1500–1900," *Journal of Global History* 5:3 (2010): 353–372; and A. C. Vigarié, "France and the Great Maritime Discoveries – Opportunities for a New Ocean Geopolicy," *Geo-Journal* 26:4 (1992): 477–481.

[23] Ibid.

[24] Ibid., 502.

[25] *Hernán Cortés: Letters from Mexico*, translated, edited and introduced by Anthony Pagden (New Haven and London, 1971), 330 and 439–440. Cortés valued the gold to be worth 3,000 gold pesos, and the jewels at 500,000. Diaz claimed that the gold was worth 58,000 Castellanos (*pesos de oro*). See Bernal Diaz del Castillo, *The History of the Conquest of New Spain*, edited and introduced by David Carrasco (Albuquerque, 2008), 323.

addition to treasure in gold, one can only imagine the wealth of information that these ships, and their now captive crews, carried. Indeed, gold, jewels, and curiosities were not the only items of value to be found on Cortés' ships – they also contained detailed reports of his exploits and information about the riches to be found in the New Worlds he had conquered. This information was found not only on paper, but also in the experience of the ships' captured crews, as for example, Alonso de Ávila, Cortés' bookkeeper, who was to be held prisoner in France for the next three years.[26]

The enormous value attributed to the market for information can be gauged both by the steady stream of detailed reports sent back to Lisbon by spies recounting Norman ventures into waters claimed by the Portuguese,[27] and by the lengths taken to prevent maps and experienced pilots from falling into the wrong hands.[28] Though not without consequence, attempts to control information were far from uniformly successful. How could they be? Rumors flew, gossip spread, charts and maps meant to be "state secrets" were traded and sold, as were the skills of translators and experienced navigators and sailors. Ango's quayside home must have been a gravitational center for the collection and distribution for both this information and the informants who trafficked in it.

A Petit-Monde

Found where the *rue du Petit-Monde* (rue d'Ango today) meets the water (present day, Quay Henri IV), Ango's home was reputed to be the finest wooden house in all of Normandy. Like his, and Parmentier's ship, it was named La Pensée.[29] According to Cotgrave, la pensée means, among other things, thinking, weighing, examining and considering.[30]

---

[26] See Guénin *Ango et ses pilotes*, 22–26, and Hernán Cortés, 329–330. Interestingly, upon his release Ávila returned to the "New World" to join in Francisco de Montejo's conquest of the Yucatán; some ten years later, he fell into the hands of the Inquisition as a relapsed Jew and a heretic, accused by Bishop Juan de Zumárraga "of keeping a crucifix under his writing desk and stepping on it." See Norman Fiering, *The Jews and the Expansion of Europe to the West, 1450 to 1800* (New York and Oxford, 2001), 188.

[27] See Wintroub, *A Savage Mirror*, chapter 2.

[28] Similar efforts were waged by the Spanish. See Jaime Cortesão, "The Pre-Columbian Discovery of America," *Geographical Journal* 89:1 (1937), 31–32; Lach, 151–154; Alison Sandman, "Controlling Knowledge: Navigation, Cartography, and Secrecy in the Early Modern Spanish Atlantic," in James Delbourgo and Nicholas Dew (eds.), *Science and Empire in the Atlantic World* (New York and London, 2008), 31–51; and Voigt, *Writing Captivity*, for example, 255–319. Regarding Spanish secrecy see María Portuondo, *Secret Science: Spanish Cosmography and the New World* (Chicago, 2009). On the limits to such strategies of information hoarding see Joan-Pau Rubiés, *Travel and Ethnology in the Renaissance: South India through European Eyes, 1250–1625* (2000), 3–4.

[29] Guénin, *Ango et ses pilotes*, 8; Gravier, *Jean Ango*, 16.

[30] Cotgrave, *A Dictionarie*, s.v. *pensée*.

The nerves channeling this particular thought extended well beyond La Pensée's quayside location, however, ranging far and wide through networks of merchants, bankers, courtiers, pilots, sailors, spies, cosmographers, looted ships, and interrogated prisoners.[31]

Ango's house made the world small, like the street upon which it sat – a *petit monde* – it was a microcosm embracing within its walls assemblages of people, things, and ideas that were dispersed far and wide across Europe and the world.[32] One can thus trace Ango's "thought" through contacts extending from the brightest lights of provincial humanist culture, to cosmographers, pirates, businessmen and royal patrons: men such as Jean Doublet and the brothers Miffant, translators of Terence, Cicero and Xenophon;[33] the mathematician and cartographer, Pierre Desceliers;[34] merchants, explorers, and cartographers such as Jean Parmentier, Jacques Cartier, Jean Rotz, and Giovanni da Verrazano; fellow investors like Zanobi and Alessandro Rucellai, cousins to the Strozzi and to the Medici;[35] pirate-explorers like Jean Fleury; and royal patrons, such as the Cardinal Georges d'Amboise, Marguerite de Navarre, and the King himself, François I, who visited La Pensée in 1534.[36] One can imagine that guests who partook in Ango's hospitality at La Pensée also included sailors, navigators, cartographers, and pilots whose names have now been forgotten, such as the Portuguese pilots of his ships, captained by Pierre Caunay and Jean Breuilly, which sailed to the Indes in 1526 and 1528.[37] Others, however, were extremely well known, as for example the famous Portuguese "turncoat" Jean Fonteneau, a.k.a. Jean Alfonse de Saintonge, who arrived in France just around the time that Parmentier embarked for Sumatra in 1529.[38] Did Ango keep prisoners from the two vessels captured by Fleury in 1524 at La Pensée? Perhaps. Whatever the case, Parmentier certainly brought – and Ango sent – Portuguese informants along with him to Sumatra. For example, Antoine, known

---

[31] Guénin, *Ango et ses pilotes*, 192.

[32] See, in particular, Mialet on the notion of the distributed centered subject; the best account of this can be found in *L'entreprise créatrice* (Paris, 2008), this might be compared and supplemented fruitfully with Foucault's notion of heterotopia, Michel Foucault, "Of Other Spaces"; how, for example, might human subjects be understood, like geographic sites, as dispersed – while centered – in extended networks?

[33] Guénin, *Ango et ses pilotes*, 20–21.

[34] On Desceliers, see Chet Van Duzer, *The World for a King: Pierre Desceliers' World Map of 1550* (London, 2015).

[35] According to Charles de La Roncière, *Histoire de la marine française*, 3 volumes (Paris, 1906), Vol. 3, 246, who overstates the closeness of the relation.

[36] Ibid., 246–247.

[37] Christian Buchet, Michel Vergé-Franceschi, *La mer, la France et l'Amérique latine* (Paris, 2006), 269–270.

[38] Who some identify as Rabelais's hero, Xenomanes. La Roncière, *Histoire*, 222–333.

as "the Portuguese," features prominently in Crignon's Log as an opinionated – and often assertive – navigator, who frequently disagreed with Parmentier's and Crignon's observations, measurements and judgments (see Chapter 5). To Antoine, we can add the expedition's Malay-speaking *truchements*, Jean Masson and Nicolas Bout, who could only have been instructed in the Malay language while in the service of the Portuguese.

La Pensée's dispersal into the world and the work it carried out as a conduit of information took place not only through the bodies of informants, but also in things that could inform, that is, through texts, instruments, and maps. As is evident from their poetry, Parmentier and Crignon were steeped in the classics; recall for example, that Parmentier dedicated his translation of Sallust's *Catiline Conspiracy* to his patron, Jean Ango, and that he was working on a translation of the *Jugurtha* for the King when he died. Ango, the Parmentiers and Crignon, moreover, surely read – and discussed – accounts of Sumatra (Taprobana) in Strabo's *Geographia*, in Pomponius Mela's *De totius orbis descriptione*, and in Ptolemy's *Geographica*, all of which were published in multiple editions in the first quarter of the sixteenth century.[39] Similarly, they must have been familiar with texts by Marco Polo, Odoric of Pordenone and Mandeville. More current accounts, of course, might also have been available. For instance, they must have attentively read the French translation of Pigafetta's account of Magellan's voyage, *Le voyage et navigation, faict par les Espaignolz es Isles de Mollucques* published in Paris by Jacques Lefèvre d'Étaples and Simon de Colines (1525 and 1526); and the account by Maximilian of Transylvania, *De Moluccis insulis atque aliis pluribus mirandis quae novissima castellanorum navigatio Sereniss*, published in 1522 and 1523 in Rome. Perhaps they also read – or heard about – manuscript copies of the *Livro de Duarte Barbosa*, Tomé Pires's *Suma Oriental*, Francisco Rodrigues's *Livro de geographia Oriental*, or the *Itinerario de Ludouico de Varthema Bolognese*, which talked about Sumatra's pepper, its gold and its perfumed woods (first published in Rome in 1510, and then again in multiple editions in Venice, Milan, Augsburg, Strasbourg, and Seville);[40] perhaps they had even seen (smelled, tasted and bought)

[39] From the late fifteenth century, and prior to Parmentier's departure in 1529, Strabo's *Geographica* was published in Venice, Basel, Treviso, Strasbourg and Paris (the 1512 edition by Lefèvre d'Étaples, for example); Mela's *De totius orbis* was similarly widely published, including an edition by Geoffroy Tory (Paris, 1507), as was Ptolemy's *Geographica*, which similarly saw multiple editions in the same span of time, for example, Waldseemüller's 1513 Strasbourg edition.

[40] See *The Travels of Ludovico di Varthema in Egypt, Syria, Arabia Deserta and Arabia Felix, in Persia, India, and Ethiopia, A.D. 1503 to 1508*, introduced and edited by G. P. Badger, and trans by J. W. Jones (London, 1863), see esp. 237–243. Regarding the edition printed in Augsburg by Hans Miller, 1515, and illustrated by Jörg Breu, see Stephanie Leitch,

samples of this pepper, as well as other spices, that were being imported in ever greater amounts to Lisbon and Antwerp.[41] In addition, maybe they also had a chance to read Martín Fernández de Enciso's *Summa de geographia* (Seville, 1519), or Andrea Corsali's epistolary accounts of his travels in 1516 and 1517 that spoke of his desire to correct – with his trusty astrolabe – the many errors in Ptolemy's understanding of the Orient's geography, such as his conflation of Ceylon and Sumatra under the name Taprobane.[42] It is perhaps worth noting that Corsali's travels were underwritten by Giuliano di Lorenzo de' Medici, cousin to the business associates with whom Ango sponsored Verrazano's 1524 voyage, the Rucellai.[43] Though Parmentier probably did not read Verrazano's manuscript account of this expedition seeing that it was written for the eyes of the king, François I, maybe he saw a draft, or a copy, or heard about it from Verrazano himself, or from sailors, navigators and pilots with whom he came into contact, whether at La Pensée, on the docks, at Church or in a Tavern.[44]

One can imagine that pamphlets, such as the Portuguese king's letter to Pope Julius II (*Epistola serenissime Regis Portugalliae de victoria contra infidels*), printed in Paris in 1507, came into La Pensée's orbit and into Ango's and Parmentier's hands. In this letter, the king recounts news of Almeida's 1505 arrival in Ceylon, including Almeida's success in reaching the "famous island of Taprobane (Sumatra)," which was, he says, considered as "another world" – *alterum aliquando orbem*.[45] Apian's gloss and emendation of Ptolemy in his *Cosmographicus* was surely a book that they had all seen;[46] perhaps they had also seen the compilations published in 1502 by

Mapping Ethnography in Early Modern Germany: New Worlds in Print Culture (New York, 2010), 103ff; see esp. Rubies, *Travel and Ethnology,* chapter 4.

[41] According to Donald F. Lach, "in the decade after 1505, Portuguese vessels annually brought into Lisbon spice cargoes averaging from 25,000 to 30,000 hundredweight [of pepper]. See *Asia in the Making of Europe*, Volume I: *The Century of Discovery,* Book 1 (Chicago, 1994); he cites as his source Gino Luzzato, *Storia economica dell'età moderna e contemporanea* (Padua, 1938), t. I: 157.

[42] See W. G. L. Randles, "La diffusion dans l'Europe du XVIᵉ siècle des connaissances géographique dues aux découvertes portugaises," in Randles, *Geography, Cartography and Nautical Science in the Renaissance* (Burlington and Hampshire, 2000), 269–277, at 275.

[43] See Bernard Beck, "Les Italiens de la mer. Marins et cartographes au service de la Normandie au XVIᵉ siècle," *Cahier des Annales de Normandie* 29 (2000). Les Italiens en Normandie, de l'étranger à l'immigré: Actes du colloque de Cerisy-la-Salle (8–11 octobre 1998): 129–142, 133–134; and Michel Mollat and Jacques Habert, *Giovanni et Girolamo Verrazano, navigateurs de François Iᵉʳ* (Paris, 1982), xx.

[44] The Manuscrit Cèllere at the John Pierpont Morgan Library, New York, ms. MA 776; and as transcribed, translated and annotated by Mollat du Jourdin and Habert, *Giovanni et Girolamo Verrazano*, 11–49.

[45] W. G. L. Randles, La diffusion, 271.

[46] With corrections and additions by Gemma Frisius (Antwerp, 1529).

Valentim Fernandes that included not only accounts by Marco Polo and Niccolò de' Conti, but the letters of the Florentine merchant, Girolamo Sernigi, who visited India, Sumatra and Ceylon in the early 1490s; or perhaps the *Paesi novamenti retrovati* by Fracanzano da Montalboddo (Vicenza, 1507), where "details on trade routes, commercial practices, and local customs" were to be found.[47] The text by Montalboddo, which included an account of Vespucci's travels, is perhaps particularly relevant as it was published in Paris in 1516 by Pierre Vidoue and Galliot Du Pré, both of whom were connected to the same Norman poetry confraternities as Crignon and Parmentier (see Chapter 4). Another letter by Sernigi, though not published, was glossed by the famous humanist collector, Konrad Peutinger. It is likely that Peutinger saw this letter thanks to Valentim Fernandes, who looked after the interests of the Augsburg merchant banking companies, the Welsers and the Fuggers, in Lisbon. In this capacity, Fernandes frequently communicated with Peutinger who was an advisor to – and a representative of – both families.[48] Peutinger, in turn, collected Fernandes' dispatches, publishing some of them as *De insulis et peregrinationibus Lusitanorum* in his *Sermones conuiuales* (Strasburg, 1506).[49]

Fernandes, however, did more than just write letters, he also helped broker the inclusion of three ships representing the interests of several German merchant houses in Francisco de Almeida's fleet bound for the East Indies in 1505.[50] Balthasar Sprenger's account of his experiences as part of this expedition was published a few years later as *Die Merfart und erfarung nüwer Schiffung und Wege zu viln unerkanten Inseln und Künigreichen* (1509); this text included important "ethnographic" woodblocks by Wolf Traut, and also, through the intervention of Peutinger, at the behest of his Welser patrons, a woodblock frieze by Hans Burgkmair that charted Sprenger's voyage through a series of encounters with native peoples as he traveled along the coast of Africa and then east to Cochin.[51]

---

[47] Lach, *Asia in the Making*, 155; also see Álvaro Velho, *A Journal of the First Voyage of Vasco Da Gama, 1497–1499* (London, 1898), 119–142. German translations by Jobst Ruchamer and Henning Ghetelin were published in 1508, while a French translation, *SEnsuyt le nouueau môde et nauigations: faictes par Emeric de vespuce Florentin* was published in Paris in 1516. See, for example, *Discovery: An Exhibition of Books Relating to the Age of Geographical Discovery and Exploration*, Prepared for the Fifth Annual Meeting of the Society for the History of Discoveries, Lilly Library, Indiana University, November 12–13, 1965 at www.indiana.edu/~liblilly/etexts/discovery/.

[48] Peutinger was married to Anton Welser's daughter, Margarethe.

[49] Leitch, *Mapping Ethnography*, 214 n.48; also see note 11, Chapter 3.

[50] Ibid., 73; see also, for example, Mark Häberlein, *The Fuggers of Augsburg: Pursuing Wealth and Honor in Renaissance Germany* (Charlottesville, 2012); and in particular, Henry Harrisse, *Americus Vespuccius: A Critical and Documentary Review of Two Recent English Books Concerning that Navigator* (London, 1895), 23–26.

[51] See Leitch, *Mapping Ethnography*, 64–73.

Burgkmair's woodcuts were printed as a pamphlet in Augsburg in 1508.[52] Sprenger's text was reissued, with pirated and reimagined images by Georg Glockendon the Elder in Nuremburg in 1511, and Burgkmair's woodblocks were cannibalized by Jan van Doesborch in a pamphlet *Die reyse vã Lissebone om te varē na d[en] eylãdt Naguaria in groot Jndien gheleghen voor bi Callicuten* (Antwerp, c. 1508), which freely mixed Sprenger's text with excerpts from Amerigo Vespucci's *Mundus novus*. A Latin broadside that included many of the same woodcuts along with texts by Sprenger and Vespucci, *De novo mondo et figura noni praecepti*, was published in c. 1509; Flemish and English versions (*Of the newe la[n]des and of ye people founde by the messengers of the kynge of porty[n]gale*) appeared in 1508 and 1520. As we will see shortly, Sprenger's words, and either Traut's, Burgkmair's, Glockendon's or van Doesborch's images, reached Dieppe, and Ango's eyes, some time in the 1520s.[53]

Maps and charts that included the routes to be followed by Le Sacre and La Pensée were also circulating. According to Donald Lach:

the Portuguese themselves prepared detailed surveys and maps for the information of the court. Although these were not published at the time, some of the new knowledge that they contained was smuggled out in the letters prepared by the informants and agents of German and Italian merchant houses. Despite the difficulty of obtaining general information on India and other parts of Asia, a hazy picture of the East could be reconstructed by those who were interested in doing so. The new materials, fragmentary as they were, could be and were used to test and supplement the accounts and maps inherited from the pre discovery era.[54]

What held for German and Italian merchant houses was surely also the case for French merchants such as Ango and houses such as La Pensée. Indeed, one can imagine that like Peutinger, who collected everything he could pertaining to Portuguese activities in the Indes, that Ango too – before launching three expeditions to the Indes orientales in as many years – similarly gathered extensive intelligence about Sumatra and the Moluccas.[55] One wonders whether copies of maps by Ribeiro, Schöener, Fine, Rodrigues, etc., were circulating in early sixteenth-century Dieppe, in addition to maps, charts and globes

---

[52] Thus preceding the woodblock prints produced by Traut.
[53] One can only surmise that out of the 1,200 men who traveled with Almeida in 1505, and the countless (because indeterminate) who followed, there must have been many letters written and accounts penned that were never published but circulated widely. For example, that of Hans Mayr, known today only through Valentim Fernandes's Portuguese translation found in a manuscript at the library of his friend and associate Konrad Peutinger. Ibid., 89. See Harrisse, *Americus Vespuccius*.
[54] Lach, *Asia in the Making*, 155.
[55] On Peutinger, see Leitch, *Mapping Ethnography*, esp. chapter 4.

by Desceliers, Cousin, Gonneville, Parmentier, and Crignon that are now lost. Certainly, members of Dieppe's merchant-maritime community had access to the 1508 Rome edition of Ptolemy's *Geographica* that included a copper-plate engraving of Johann Ruysch's world map reflecting information culled from recent Portuguese explorations of the Indian Ocean. Or perhaps they saw the 1505 "newsletter" *Den rechte[n]weg ausz zu faren von Liszbona gen Kallakuth vo[n] meyl zu meyl*, that included a map of the newly found route to the East. And surely, copies of the emended Ptolemaic world atlases by Waldseemüller and Laurent Fries, published in Strasbourg, were to be found in Normandy; these included references to recent voyages by the Spanish and the Portuguese. Were maps constructed, or intelligence gathered, by the likes of António de Abreu, Francisco Serrão or Francisco Rodrigues on their 1512 expedition to the spice islands also clandestinely circulating? Perhaps Ango and his men had also heard about, or seen, copies or sketches of maps or charts by Pedro and Jorge Reinel and Lopo Homem. All of the surviving maps associated with the "Dieppe school of cartography" – including, for instance, maps by Jean Rotz, Pierre Desceliers, Nicolas Desliens, and Guillaume Le Testu – had place names either in Portuguese or in Gallicized Portuguese, leaving little doubt that the Dieppois mapmakers associated with Ango had access to – and depended upon – Portuguese charts and/or captured or bribed informants.[56] Indeed, we know by internal references in Crignon's Log, that he used such a chart in guiding La Pensée to Sumatra. Thus the entry for Wednesday, June 16, notes:

the height [of the sun] was taken at noon: 35 degrees east (*l'orient*) to 40 degrees west, the south to 17 degrees. And the longitude east, at the point on the map was marked with an 'A,' and the point of the longitude [west] was indicated on the map by a 'V.' The wind that day was good and to the west; the cape [of Good Hope] was east-southeast.[57]

Where might such a precise map – one that even included longitudes off the south-eastern coast of Africa – have come from? In one way or another, it must have originated with the Portuguese.[58]

---

[56] See, for example, Sarah Toulouse, "Marine Cartography and Navigation in Renaissance France," in David Woodward (ed.), *The History of Cartography*, Vol. 3 (Chicago, 2007), 1550–1568; on Dieppe's mapmakers, see also S. Davies, *Renaissance Ethnography and the Invention of the Human: New Worlds, Maps and Monsters* (Cambridge, 2016).

[57] Crignon, *Oeuvres*, 22.

[58] It was, however, likely to have been a French copy, or dated, as its information seems to have been either wrong or inconclusive in the disagreement that would break out a few months later between Antoine the Portuguese and Jean Parmentier over the location of the Maldives. See Chapter 5.

Le Petit Monde **and the Ocean Seas**

Wherever and however information was accessed, it was never simply a thing in itself – unmediated, raw or unproblematic; rather, it traveled in connected chains of reactive feedback, each iteration, each link, mutating in relation to (while also attempting to stabilize) new voyages, new questions, diverse and conflicting interpretations and complex interactions with new informants, new maps, new instruments and new – and always changing – experiences. The "cycles of accumulation" evinced by "centers of calculation" such as Peutinger's library, and conjecturally, by Ango's La Pensée, were moreover integrated into the development of what, anachronistically, might be termed strategic planning, risk management, and instruments of credit.[59] The house on Dieppe's *rue du Petit-Monde* was not simply an architectural end point of information (in the modern sense); it was an *escale* (a port) belonging to any number of circulating voyages. Similarly, the information that circulated through it was not discrete decontextualized data any more than the house that contained it, but knowledge enchained to mobile assemblages of people, ships, instruments, and texts. La Pensée – as both ship and home – was, in this sense, a kind of wooden translating machine that was harnessed to the mobility of information, and as such, to quests for souls, spice, status and profit.[60] La Pensée was, quite literally, a heterotopia: a place that was another place, like a ship sailing – a thought running – to the farthest reaches of the earth. Thus, though a wooden structure found on the quay next to *rue du Petit-Monde* in Dieppe, Ango's home was also an assemblage of artifacts, instruments, texts, and people: it was spice traders in Antwerp; travel books printed in Augsburg; instruments made in Paris; maps constructed in Lisbon; pirates and poets from Dieppe; texts written by the ancients; spies and agents placed among allies and competitors; ships, like her namesake, sailing in distant seas; and Minangkabou spice merchants on the north central coast of Sumatra.

[59] The prediction of future results, which is to say, the normalization of transoceanic shipping and trade, could only be performed in relation to the kind, and the quality, of information available; this was true whether for the judgment of less formal associations as the financial backers of Verazzano's search for a Northwest Passage in 1524, or for more formalized institutional mechanisms for risk management employed in maritime insurance. See Chapter 5.

[60] In many regards, it might fruitfully be compared to a curiosity cabinet – a place of wonder and exoticism that made available the world in microcosm. Cabinets of curiosity, like Ango's quayside home, were far from self-contained and bounded hermetically within themselves, they were similarly entangled in extended and far-reaching networks. See, for example, Christian Feest, "The People of Calicut: Objects, Texts, and Images in the Age of Proto-Ethnography," *Bol. Mus. Para. Emílio Goeldi. Cienc. Hum.*, *Belém*, 9:2 (2014): 287–303, 291.

Though a heterotopia, La Pensée was also a physical place: a singular site built of wood and stone that on July 22, 1694 found itself in the crosshairs of an Anglo-Dutch bombardment that would destroy it forever, except for a few paving stones. From heterotopia to utopia – from many places to none – to literally "no place," it was gone with hardly a trace, for along with La Pensée, all of Dieppe's archives were also destroyed by Lord Berkeley's fleet.

There was, however, a more durable heterotopia in sixteenth-century Dieppe that survived the Anglo-Dutch assault. It was not La Pensée, to be sure, but it was a vessel, a sacred and consecrated site constructed to transport the souls of the living across material and spiritual divides: L'église Saint-Jacques, where Jean Parmentier and Pierre Crignon read their poetry each year on August 15 for the Puy de l'Assomption and where, in 1551, Jean Ango found his final resting place.

Saint-Jacques and the World

Built, in stops and starts, over the course of four centuries, it has been suggested that l'église Saint-Jacques was named after this saint because, like the Dieppois, he was a man of the sea;[61] however, it was more likely that the church owes its origins to the initiative of the archbishop of Rouen and his desire to link his diocese to popular – and very profitable – pilgrimage routes, than to fishermen in Dieppe. According to legend, before his martyrdom at the hands of Herod Agrippa, Saint-Jacques (Santiago) had evangelized Iberia.[62] After his death in Judea, his relics were translated back to Galicia guided by angels on a rudderless unmanned boat. They were then transported to Compostela where they were forgotten until the early ninth century, when they were miraculously "rediscovered."

Though we tend to think of the cult of Saint Jacques as a Spanish phenomenon, the French were among its most enthusiastic boosters.[63] Clear references to the popularity of Saint-Jacques along the northern French

---

[61] David Asseline, *Les antiquitez et chroniques de la ville de Dieppe,… publiées pour la première fois avec une introduction et des notes historiques par MM. Michel Hardy, Guérillon et l'abbé Sauvage* (Dieppe, 1874), 122.

[62] See for example, Jan van Herwaarden, "The Origins of the Cult of St. James of Compostela," *Journal of Medieval History* 6 (1980): 1–35. This narrative is told, for example, by the beautiful mid sixteenth-century windows (1548) at the Chapelle Notre Dame du Cran in Brittany.

[63] See Asseline, *Les antiquitez*, 123, and, for example, Marcelin Defourneaux, *Les Français en Espagne aux XIe et XIIe siècles* (Paris, 1949); Denise Péricatd-Méa, *Compostelle et cultes de Saint-Jacques au Moyen Âge* (Paris, 2000); Adeline Rucquoi (ed.), *Saint-Jacques et la France: actes du colloque des 18 et 19 janvier 2001 à la fondation Singer-Polignac* (Paris, 2003).

coast can be found in several Norman churches in and around Dieppe. For example, in addition to l'église Saint-Jacques in Dieppe, there was another medieval église Saint-Jacques in Tréport just to the north. Inland from Dieppe, in the Andelys, was the thirteenth-century hospice Saint-Jacques charged with the care of pilgrims in transit to Compostela, while just south was l'église Saint-Jacques in Lisieux with its sixteenth-century windows depicting the saint's life. Both Rouen's Cathedral, and l'Abbé Saint-Ouen, prominently featured images and sculptures of the Saint as Peregrino. Scallop shells – symbol of the saint and his pilgrimage – were decoratively sculpted into interior pillars in l'église Saint-Valery in Varengeville-sur-Mer, and in l'église Saint Martin in Veules-les-Roses, just south of Dieppe. The enduring popularity of Saint-Jacques and the pilgrimage route leading from Dieppe to Compostela can perhaps be gauged by the seventeenth-century shell sculptured into the base of the house belonging to the descendants of Ango's business associates, the Miffants, on Dieppe's rue d'Ecosse.

Among the first – and certainly among the most popular – accounts of the Camino was a collection of texts by a French monk from Poitou, Aymeric Picaud, known as the *Codex Calixtinus* (c. 1135–9).[64] This included the *Historia Caroli Magni et Rotholandi* that purported to be an eyewitness account of Charlemagne's ninth-century assault on the Moors of Spain, as well as the *Iter pro peregrinis ad Compostellam*, an extremely popular "tourist" guide and proto-ethnographic travel account that detailed not only what a (French) pilgrim to Santiago's shrine ought to see, do, and avoid, but also descriptions of the cultural characteristics of the different peoples that he was likely to meet along the way.[65]

The *Pseudo-Turpin Chronicle* was allegedly authored by the eighth-century Archbishop of Reims, Turpin. The monk and ecclesiastic, however, was here supplanted by a fictional Turpin, one of the 12 paladins of Charlemagne, whose heroic deeds performed against the Saracens can be found in the literary cycle of the *chanson de gestes*, most notably, the *Chanson de Roland*. Internal evidence in the Chronicle clearly points to its twelfth-century origins by an anonymous French author who was surely either a contemporary of the authors who wrote the *chansons de geste*, or was familiar with the oral tradition that preceded their written form. The

---

[64] Well over 100 manuscript copies have survived. See Stephen G. Nichols, *Romanesque Signs: Early Medieval Narrative and Iconography* (New Haven, 1983), 127. On the codex's French provenance see H. M. Smyser (trans. and ed.), *The Pseudo-Turpin* (Cambridge, MA, 1937), 3.

[65] The people of Gascon, for example, were characterized as "fast-talking, obnoxious, and sex-crazed, they are overfed, poorly-dressed drunks. They've two good characteristics: they are skilled warriors, and they give good hospitality to the poor." See Chapter VII: http://codexcalixtinus.es/the-english-version-of-the-book-v-codex-calixtinus/

Psuedo-Turpin is particularly important in the present context as it was closely associated both with the development of an enduring literary tradition of sacred (con)quest, and the popular image of Saint-Jacques as a proto-nationalist Christian conqueror. The chronicle, also known as the *Historia Caroli Magni*, recounts how Saint-Jacques visited Charlemagne in his dreams to exhort him, "as the great liberator of Christianity, to earn the crown of eternal blessedness by faring forth under the way of stars against the pagan Galicians."[66] Charlemagne, as directed by the Saint, thus followed the path of stars from France across the Pyrenees to Galicia to battle for the faithful against Saracens and Moors, and win back and make safe for the pilgrims that were to come – the forgotten place, Compostela, where the saint's body had been put to rest close to a millennium before. Not just a dream, however, Saint-Jacques also appeared above the ramparts of Muslim-occupied Pamplona, miraculously causing them to crumble, thus bringing Charlemagne's siege of the city to a decisive end.[67] The rest of the peninsula was quickly liberated; Charlemagne baptized those who were willing and killed those who were not. He then built and embellished the shrine at Compostela with the booty pillaged and tribute exacted from the Muslims he had conquered. After returning to France, the Chronicle recounts how Charlemagne founded numerous churches devoted to Saint-Jacques, and how he tirelessly promoted pilgrimage to his shrine.[68] Charlemagne was, in this sense, both a conquering liberator of Iberia, and Saint-Jacques' first pilgrim.

Like the *Chanson de Roland*, the *Psuedo-Turpin Chronicle* was a twelfth-century fiction, roughly contemporaneous with Spanish accounts of Saint-Jacques' miraculous appearance at the mythical ninth-century Battle of Clavijo. According to the self-serving forgery by the Compostelan canon, Pedro Marcio, known as the *Diploma of Ramiro I*, Saint-Jacques (Santiago) appeared at Clavijo riding a white horse and wielding a bloody sword to lead the outnumbered and beleaguered Christian armies to triumph over a superior Muslim host. As the Matamoros, the Moor slayer, Saint-Jacques was the prototype of the crusading Christian warrior.[69] Iconographically, he was represented sporting his ubiquitous icon, the coquille St. Jacques, riding a battle horse and swinging a sword. Beginning in the fifteenth century, he was depicted triumphantly trampling over the mutilated bodies of the Muslims he had slain. This bloodthirsty representation of

---

[66] Smyser, *The Pseudo-Turpin*, 18.
[67] Ibid.
[68] Ibid., 21–22.
[69] The image of the crusading saint presented by these texts was closely associated with Cluny; ibid., 3–4. See also, Émile Mâle, *L'art religieux du XIIe siècle en France: étude sur les origines de l'iconographie du moyen âge* (Paris, 1922), 292.

Saint-Jacques as the Matamoros seems entirely distinct from the much more common, and pacific, iconography of the saint as a barefooted *Peregrino*, with a pilgrim's staff, a traveller's bag, a cap – typically graced with a scallop shell – and, sometimes, a Bible, a reference to his apostolic role in the conversion of Iberia. It is clear, however, that despite these very different iconographic programs, the two St. Jacques were very similar: pilgrimage, like crusade, was a conquest and pacification of space. Staff and sword were, in this regard, interchangeable attributes that accompanied movement across physical and spiritual distance. Santiago Peregrino and Santiago Matamoros were complementary rather than antithetical – they were two sides of the same impulse toward Christian imperium.[70] Put somewhat differently, like Charlemagne, his human avatar, Saint-Jacques was both pilgrim and conquering warrior; similarly, while the act of pilgrimage, which he represented as Peregrino, was a spiritual voyage, it was also – as its Cluniac propagandists intended – holy war, crusade and (re)conquest. It is thus perhaps not surprising, given this history, that Santiago Matamoros would become for sixteenth-century conquistadors in the New World, Santiago Mataindios – the Indian slayer – or that men like Vasco da Gama, Francisco de Almeida, and Afonso de Albuquerque were all members of the Order of Santiago.[71]

[70] Stephen B. Raulston, "The Harmony of Staff and Sword: How Medieval Thinkers Saw Santiago Peregrino and Matamoros," *La Corónica: A Journal of Medieval Hispanic Languages, Literatures, and Cultures* 36:2 (2008): 345–367. See also, in the same volume, John K. Moore, Jr., "Juxtaposing James the Greater: Interpreting the Interstices of Santiago as Peregrino and Matamoros," 313–344; and J. K. Moore, "Santiago's Sinister Hand: Hybrid Identity in the Statue of Saint James the Greater at Santa Marta de Tera," *Peregrinations: Journal of Medieval Art & Architecture* 4:3 (2014): 31–62. I would like to thank John Moore for generously sharing his thoughts on the two Santiagos with me.

[71] Contemporaneous with the writing and collection of these texts about Saint-Jacques, was the emergence of the knightly order of Santiago. According to its founding charter (1175), the order was charged with the defense of Christendom against infidels and Moors. The role played by the order was, of course, not so clear cut as its imperial charter would have one believe; indeed, the order was different things to different groups – whether Castilian or Portuguese, north or south, southern Portuguese nobles, or crusading conquistadors, or indeed, the monarchs who sponsored them. Sanjay Subrahmanyam has admirably discussed some of these complexities; for our purposes, it is important to note that whilst conquistadors were employing "Santiago" as a battle cry in their wars of conquest in the New World, that Portuguese explorers, colonists, and administrators had similarly embarked on crusading missions in Africa and the Indes orientales in the name of Santiago. Sanjay Subrahmanyam, *The Career and Legend of Vasco Da Gama* (Cambridge, 1998), 22–73. See Luís Adão da Fonseca, "The Portuguese Military Orders and the Oceanic Navigations: From Piracy to Empire (Fifteenth to Early Sixteenth Centuries), in M. Barber and J. M. Upton-Ward (eds.), *The Military Orders: On Land and By Sea* (Hampshire, 2008), 63–76. In Spain, for example, members of the Order included Hernán Cortés, Hernando de Soto, and Menendez de Aviles (who was responsible for eliminating the Protestant French colony, including its leader from Dieppe, Jean Ribaut, in Florida in the 1560s).

The Charlemagne pilgrimage-quest as recounted by the *Psuedo-Turpin Chronicle* and subsequently translated into the *chansons de geste* cycle, along with the continued vitality of the pilgrimage tradition it inaugurated and sustained, speaks to the foundational importance of Saint-Jacques not only for the Spanish and the Portuguese, but for the French. Indeed, of note here, is not only the fact that Dieppe's Church of Saint-Jacques was a stop along the way of a well established and popular pilgrimage route, but that this pilgrimage was entangled in traditions that valorized both Saint-Jacques' apostolic role in the evangelization of Galicia through the extension of God's Word (Peregrino) and his role as a conquering warrior and killer of heathens (*Matamoros* and *Mataindios*). This dual figuration of Saint-Jacques as pilgrim-conqueror was widely represented in France as well as in the Iberian peninsula in the sixteenth century, as demonstrated by elaborate church windows such as those found in the église Saint-Pantaléon (c. 1539) in Troyes and also in Notre-Dame-en-Vaux in Châlons-en-Champagne (c. 1525) (see Figure 1.1). Information, taken in this sense, is far more than just intelligence about faraway places; it is also generative of quests and imperial adventures, such as those carried out by the progenitor of French kings, Charlemagne, against Saracen invaders of Iberia, and many centuries later by Ango and his pirate-privateers, who worshipped, prayed, and read their poetry in Saint-Jacques' nave.

Pilgrimage Architecture

L'église Saint-Jacques was a ship of stone and glass built and designed to carry its supplicants on a spiritual quest. Jacques was, after all, a sailor and a fisherman, as well as a warrior and an apostolic pilgrim, but in this case, his ship was a church, *un nef*, French for both ship and nave (see Chapter 3). Saint-Jacques's resemblance to a ship (see Figure 1.2) was certainly not an accident. Even the church's cruciform shape – generalized from the early years of Christendom – with the transept bisecting the *nef* at the eastern end of its longitudinal axis, was a commonplace reference to a ship put to sail; a cross to be sure, but also a mainmast and yard.[72]

Saint-Jacques was a heterotopian space constructed as a stationary voyage within the nexus of the church and the supplicant's soul; at the same time, it was a port-of-call – an *escale* – on a pilgrimage along the

---

[72] L'église Saint-Jacques of Dieppe was not unique in this regard; churches were commonly built in the form of a cross, and were, moreover, closely associated with ships that would carry humanity toward salvation. This is an ancient association, tracing itself to the biblical precedent of Noah's ark and to the gospel according to Matthew.

Figure 1.1 Saint-Jacques as the Matamoros, stained-glass windows at Notre-Dame-en-Vaux, Châlons-en-Champagne (c. 1525).

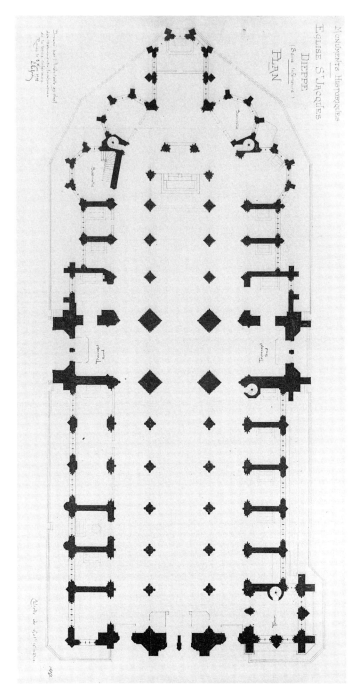

Figure 1.2 Plan of l'eglise Saint-Jacques, Dieppe. Lucien Lefort (1912).

sacred itinerary of the *Route des Anglais* to Compostela. In other words, l'église Saint-Jacques was not only a stop along the way of a well-traveled pilgrimage route, it was also a meditative conduit, a place where the sensorium subsisting between passenger and church could substitute for, recapitulate and animate a spiritual voyage. In this sense, the Church was like a text recounting a voyage – a travel account or a narrative of a pilgrimage – that was defined by a procession of words that formed a picture through a detailed circumstantial description of places and sights that needed to be experienced as stages, steps, and ports-of-call narrating movement toward a sacred endpoint. Eyes were to traverse pages, like legs striding over ground, in an ineffable and inexorable sequential progress – a quest – toward a spiritual union with the divine.[73] Texts were, in this sense, witnesses testifying to the reality of an allegorical topography of pilgrimage in much the same way as the artistic and symbolic instruments that organized voyages made possible by churches such as Saint-Jacques.

Histories of quests and voyages to faraway worlds were extremely popular; they were late medieval and early modern bestsellers. There were close thematic and literary homologies between pilgrim tales, chivalrous romances, travel accounts, and of course, *chansons de geste*. The itineraries they described were typically laced with marvels, wonders, and curiosities. These were spurs to keep reading: tantalizing morsels – *effets de présence* – that would propel readers through a text. Yet, without the spiritual-aristocratic inflection of the quest, such wonders and curiosities were thought to be dangerous, perhaps even sinister traps. Indeed, they were often associated with the lower social orders – with ignorant rustics incapable of seeing beyond the supposed proof of (sensational) experience to the essence of phenomena and true causes (*scientia*). Despite these negative connotations, the marvelous could also serve, as Ficino, for example, believed, as "signposts to … the fundamental metaphysical structure of the universe."[74] Marvels and wonders were not just good marketing tools that could boost the signal of information's circulation, they were also propaedeutic to information (that is, to the formation of Christian subjects) insofar as they provided a conduit from the particularities of the contingent world to the universal truths of philosophy and theology.[75] Albertus Magnus, for example, describes the critical

---

[73] Not unlike Lancelot's hunt for the grail. See Marc Shell, *Art and Money* (Chicago, 1995).

[74] Quoted in L. Daston and K. Park, *Wonders and the Order of Nature: 1150–1750* (New York, 1998), 161.

[75] Felipe Fernández-Armesto, *Amerigo: The Man Who Gave His Name to America* (New York, 2008), 96, 95–100.

role wonder plays in the formation of the knowing (informed) subject. Wonder, he said,

is defined as a constriction and suspension of the heart caused by amazement at the sensible appearance of something so portentous, great, and unusual, that the heart suffers a systole. Hence wonder is something like fear in its effect on the heart. This effect of wonder, then, this constriction ... springs from an unfulfilled but felt desire to know that cause of that which appears portentous and unusual: so it was in the beginning when men, up to that time unskilled, began to philosophize ... Now the man who is puzzled and wonders apparently does not know. Hence wonder is the movement of the man who does not know on his way to finding out, to get at the bottom of that at which he wonders and to determine its cause ... such is the origin of philosophy.[76]

Thus, wonders could serve as meditative and conversational loci – sources that could direct eyes and minds toward theological, moral, and natural knowledge. Erasmus's *Convivium religiousum* (*The Godly Feast*, c. 1522) describes the use of wonders in this way by locating experience of them in the didactic architecture of a garden decorated with inscriptions, sculptures, paintings, and all manner of plants. Like relics on the itinerary of a pilgrim, these artifacts – both man-made and natural – were collected and deployed to motivate and provide content to spiritual meditation as well as learned conversation. In this manner, Erasmus expressly translated the sensuous experience of collective contemplation organized by his colloquy's fictional garden into the aristocratic notion of the hunt (of inquest) – not of "stags and boars," as he put it, but of human subjects (souls) that were to be divinely (in)formed and infused with knowledge of God.[77]

Among the many models available for Erasmus's *Godly Feast* might have been the *Orti Oricellari* (the Rucellai garden) of Florence that was built and organized by Bernardo Rucellai, cousin of Pietro, Zanobi, Mario and Alessandro Rucellai, Ango's Rouennais-based business partners, thus bringing the Florentine gardens and the convivial feasts held there into the orbit of Ango's extended "informational" networks.[78] Bernardo Rucellai – brother-in-law to Lorenzo de Medici and grandson of Palla di

---

[76] Quoted in Stephen Greenblatt's *Marvelous Possessions: The Wonder of the New World* (Chicago, 2008), 81. Also see, for example, Michael McKeon *The Origins of the English Novel: 1600–1740* (Baltimore, 2002), and in particular his discussion of the trope of "strange therefore true."

[77] Erasmus, *The Collected Works of Erasmus, Colloquies*, trans. Craig Thompson, Vol. 1 (Toronto, 1997), 207.

[78] It is interesting to note, in the present context, that the shield of the Rucellai is a ship, "the mast of which is formed by a nude woman who holds in her raised left hand the mainyard and in her right the lower part of the swelling sail." See Felix Gilbert, "Bernardo Rucellai and the Orti Oricellari: A Study on the Origin of Modern Political Thought," *Journal of the Warburg and Courtauld Institutes*, 12 (1949): 101–131, 103.

Onofrio Strozzi – was a renowned collector of, and writer about, antiquities; he was also an enormously wealthy banker, and a respected historian who Erasmus praised as a new Sallust – the Roman historian that Jean Parmentier was to translate while sailing to Sumatra in 1529 (see Chapter 3).[79] The Rucellai gardens, and its feasts, had close filiations with Florence's Platonic Academy, and with such luminaries as Marsilio Ficino (of whom Rucellai was said to be a devoted disciple),[80] and Leon Battista Alberti who provided the design for the famous Rucellai *Palazzo*. The garden consisted of a collection of rare plants well known to the ancients as evinced by their presence in classical literature, and decorated with antiquities, statues, medallions, fountains and inscriptions – objects, examples, symbols and sights that could carry observers and interlocutors to other worlds and other times.

The gatherings that assembled at the Orti Oricellari in the early years of the sixteenth century are best known to us today by the participation of Machiavelli who composed his *Art of War* (1521) in its bucolic settings. Objects and plants in the gardens were explicitly meant to focalize philosophical, political, historical, literary, and religious conversations. Machiavelli, for example, comments in his *Arte della guerra* that after their feast he and his companions (Fabrizio Colonna, Zanobi Buondelmonti, Battista Della Palla, Luigi Alamanni, and Bernardo's grandson, Cosimino Rucellai) retreated to the tranquility of the garden to continue their conversation. There, in the sight of certain trees, now rare, but well known to the ancients, he was inspired to remember "several princes of the Kingdom" who in the distant past took "delight in these ancient growths and shades." This, in turn, provoked further associations with – and discussion about – the ways of the ancients and how best to imitate them.[81] Discussions ranged from history and politics, to the virtues of the Tuscan language.

Like the Puy de l'Assomption, which met each year at l'église Saint-Jacques, the humanist scholars who met at the Rucellai gardens were closely associated with the valorization of the vernacular.[82] Giovanni Battista Gelli, for example, traced "the regeneration of the Florentine

---

[79] Ibid., 102.

[80] Ibid., 104. See also, Armand L. De Gaetano, "The Florentine Academy and the Advancement of Learning Through the Vernacular: the Orti Oricellari and the Sacra Accademia," *Bibliothèque d'Humanisme et Renaissance* 30:1 (1968): 19–52, 22.

[81] Niccolò Machiavelli, *Art of War* (Chicago, 2009), 10–11. See also, Anthony Cummings, *The Maecenas and the Madrigalist: Patrons, Patronage, and the Origins of the Italian Madrigal* (Philadelphia, 2004), 15–78.

[82] See, Wintroub, *A Savage Mirror*, regarding the relationship between the Norman poetry associations and the vernacular, and Chapter 2. On the *Orti* and the promotion of the vernacular, see De Gaetano, "The Florentine Academy," *passim*.

language ... [to] the work of the Orti."[83] Indeed, Bernardo Rucellai famously refused to speak Latin when he met Erasmus in 1511, only speaking Italian, of which Erasmus had no knowledge. The Orti was also associated with the empirical investigation of curiosities, as for example its public dissections of human monstrosities (in 1536) described by Benedetto Varchi in his *On the Generation of Monsters* (this is perhaps particularly noteworthy given Parmentier's shipboard postmortem dissections of his crew discussed in Chapters 3 and 4).[84]

Like the convivial feasts and meandering walks associated with the Orti Oricellari, or its Christianized analogue in Erasmus' text, the *Convivium religiousum*, l'église Saint-Jacques comprised a journey that didactically enfolded its voyageurs in the circulation of information, disciplining their minds and organizing their bodies and eyes as they moved through its doors into the various chapels punctuating its ambulatory. Everywhere one stood, kneeled, looked, listened, and smelled there were affective and cognitive clues to be pondered and mediated upon: the incense in the air, the peal of bells, the timbre of the priest's voice, the illuminated windows, the saintly statues, the relics in reliquaries and the fragments of text written on walls and in colored light; images, paintings and sculptures, sights, sounds and smells, all designed to motivate, direct, and inscribe – to form and inform – Saint-Jacques' good Christian passengers through the body-to-body emotional contagion of collective worship. The church and the people in it were thus literally enjoined to perform a pilgrimage – they moved without moving: a microcosmic journey to other worlds.

Of course, this was a social journey every bit as much as it was a spiritual one; in the fourteenth century, for example, important members of Dieppe's civic elite, such as Guillaume de Longueil (grenétier du sel in the fourteenth century as Ango was in the sixteenth), and Baudouin Eudes (also, like Ango, an armateur), helped build chapels in honor of Notre Dame and the Holy Trinity. Many other chapels were built by the bourgeois of Dieppe in the fourteenth and fifteenth centuries,[85]

---

[83] Ibid., 25. Many of the participants in the discussions at the Orti later became members of the Accademia Fiorentina, founded in 1540, which was dedicated to the promotion of the Tuscan language. See, for example, Michael Sherberg, "The Accademia Fiorentina and the Question of the Language: The Politics of Theory in Ducal Florence," *Renaissance Quarterly* 56:1 (2003): 26–55,

[84] *Sopra la generazione de' Mostri & se sono intesi dalla Natura, o no* (1548).

[85] Guilds and confraternities also contributed, as for example, the brouettiers, les cordonniers, les savetiers-carreleurs, les chandeliers-épiciers, les cuisiniers, les bouchers, les drapiers, les merciers, les tonneliers, les canonniers, les chirurgiens, who invested in the church's sacred itinerary. For others that contributed to the building the church's chapels see A. Legris, *L'eglise Saint-Jacques de Dieppe: notice historique & descriptive* (Dieppe, 1918), 13 n. 1ff. In much the same way that ritual processions were organized in relation to status competition over proximity to the numinous – whether a saint's relic, the

culminating (for our purposes), in the sixteenth, with Jean Ango making several contributions to the Church's interior design and architecture: a personal chapel, an oratory, and the exterior façade of the trésor (about which, much more in pages to come). As we will see, these architectural ports-of-call were, at one and the same time, performative of social prestige and of the on-going instruction – the induction – of fellow-travelers in a common quest.

Saint-Jacques was, in this sense, a vessel that carried its passengers on a social and a spiritual voyage, but it was also an invitation, a goad, and a stimulus to make a physical voyage – to undertake a pilgrimage and set off into the world to carry the word, to touch and/or conquer unbelievers, and to expand Christian ideals and Christian power across geographical and cultural expanses – that is, to make a quest. This is nowhere better demonstrated than by the design of the trésor at the entrance of the sacristy on the northern side of the nave, and in particular its *"Frise des Sauvages."*

Chains of Being and Chains of Words

The "Frise des sauvages" was constructed as a chain of being carved in stone stretched in motion across the second ambulatory chapel on the northwest side of Saint-Jacques' nave. It dates from the first third of the sixteenth century, right around the time Parmentier left, or Crignon returned (see Figure 1.3). It depicts peoples from various parts of the world – Brazil, Africa, and the Indes – as well as monkeys, birds, and trees. The frieze begins with movement, a gesture from a European leaning in; this touches off a chain of motion that reverberates across the wall. We begin with a Tupinamba warrior and his king, and perhaps the king's heir, a small child, playfully climbing up the savage monarch's royal scepter. These figures are linked to those that follow by the counterpoise of weight and posture of an "African" child. His arm reaches out to the Tupinamba before him, while his face and foot are acrobatically turned to his parents behind.[86] His "father," armed with a spear, takes aim, either at a serpent entwined around the trunk of a tree, or at an owl perched on a branch above. On the other side of the tree, his mother sits, suckling an infant. The bird seems startled by the

---

Eucharist or a visiting dignitary, a prelate, a bishop or a king. See Wintroub, *A Savage Mirror*, esp. chapter 7, and idem, "Taking a Bow in the Theater of Things," *Isis* 101:4 (December 2010): 779–793.

[86] I am assuming that the frieze depicts a movement from left to right (from the viewer's perspective), across the western side of the nef from south to north, toward the church's altar.

Figure 1.3 Anonymous, "Frise des Sauvages," sculpted in stone, l'église Saint-Jacques, Dieppe (c. 1525–1535).

prospect of violence, and appears ready to take off into the frieze's next scene. Bordered by another tree, a turbaned "Arabian" man wearing a shawl reaches for the extended hand of a child; the child, however, follows the frieze forward with his gaze toward his mother; she holds him by the hand, but seems to be looking at the man. She is wearing a vale over her head that hangs to her knees. In the next scene, a Tupi wearing only a feather headdress mirrors the Tupi king at the frieze's beginning; he holds the trunk of a tree in one hand (instead of a scepter) and a palm frond in the other. The frieze continues with three men (from "Cochin") with carved shields built into a raised rectangular column that climbs up the wall; they are naked, but wear elaborate headdresses. Next to them, a turbaned man, dressed similar to the Arab who came before, looks back while blowing on what might be a trumpet, the frieze is damaged at this point, and it is no longer there.[87] The forward momentum of the frieze picks up speed here. Separated by two monkeys and a tree, a captive is led in chains behind musicians marching with a palanquin carrying a small child. The place of the shield-bearing men of Cochin thus becomes clear; they are the rear guard of a triumphal procession announced by the sound of the Arab's trumpet. A stern-looking man, armed with a very European-looking halberd, guards the litter; he is preceded by a group of three musicians. Their forward motion, however, is blocked by a Tupi couple; to their right, a drummer looks on at their naked embrace. They seem happy and content. The frieze ends with another naked Tupi with a feathered headdress; in his left hand he holds a spear or arrow, while with his right, he gestures toward three primitive men, or perhaps, as Vitet opines, orangutans. The first of these, body turned – and arms and legs – stretched out toward the Tupi, stares disconsolately out from the frieze; the men (the beasts?) behind him, sit back to back. The first bows his head toward all that came before – or who have yet to arrive, while the second turns into the wall, clasping his face in his hands, thus ending the Frieze's procession.

Despite its detailed representation of different exotic peoples visited by the Dieppois in the first quarter of the sixteenth century, the frieze was not (for the most part) constructed from the "raw material" of direct observation, as many commentators have maintained.[88] This

[87] With this exception, the frieze seems to be relatively well preserved, especially compared to the deliberate mutilations of the wall design below it, and the natural degradation of Ango's private chapel on the other side of the Church.

[88] For example, Ludovic Vitet, *Histoire de Dieppe* (Paris, 1844), 260ff. Among the many opportunities to view "exotic" peoples, one can cite, for example, Paulmier de Gonneville who returned from the "Indies Meridionale" in 1505 to Honfleur with a native named

is not to say that it was not an attempt to consecrate the (aspirational) reach of Ango's merchant empire, but rather, that the information upon which it relied was anything but "raw."[89] Indeed, the exotic peoples represented in stone in l'église Saint-Jacques were the fruit of highly mediated accounts translated into print, rather than from direct – first-hand – experience.[90] In this regard, texts from Antwerp, Nuremburg, Augsburg and Basel more greatly influenced the design and content of the frieze, and the conveyance of information about newly discovered lands, than either intelligence flowing from the Iberian peninsula or from the experienced voyagers who sailed on Ango's behalf to Africa, India, and America. Indeed, the individual vignettes of native life that compose the frieze's itinerary have clear references to a thriving culture of print in early modern Normandy. While much of this, as we will see, was Northern European in origin, it also had important Italian antecedents, both directly and indirectly.

The overall effect of the frieze is motion – movement materialized in stone. The voyage undertaken by the observer's eye of exotic peoples carved high up onto the *nef* is paralleled by the momentum of the frieze carried by its representation of a triumphal march – literally, a "progress" across the wall. The relative frame demarcated by individual bodies represented in contrapposto, with weight shifting, turning, twisting and balancing, as well as marching, further complicates the frieze's illusion of a flowing series. There are any number of important antecedents to this triumphal theme; these include, of course, Mantegna's *Triumph of Caesar*, but also the twelve woodcuts it inspired by Benedetto Bordone in 1503–4, and copies and interpretations

---

Essomericq. A few years later, and closer to home, Thomas Aubert, in the employ of Jean Ango's father, and sailing either *La Pensée* or its namesake, returned from Terre Neuve with seven natives, today identified as Micmacs; while some 20 years later, Le Sacre and La Pensée returned from Sumatra with six natives that the Portuguese had abandoned off the coast of Africa (see Chapter 6). Indeed, from August and December of 1529, the same year that Parmentier and Crignon set off for Sumatra, not les than four ships returned to Honfleur from "Brazil" (that is, from *France Antartique*). See P. Gaffarel, "Anciens Voyages Normands au Brésil," the *Bulletin de la Société de l'histoire de Normandie* 5 (1887–1890): 236–239; and E. Gosselin, *Documents authentiques et inédits pour servir à l'histoire de la marine Normand* (Rouen, 1876), 142–171. See also, for example, Vincent Masse, "Les 'sept hommes sauvages' de 1509: Fortune éditoriale de la première séquelle imprimée des contacts franco-américains," in Andreas Motsch and Grégoire Holtz (eds.), *Éditer la Nouvelle-France* (Laval, 2011), 83–106, and Wintroub, *Savage Mirror*, 23.

[89] See Chapter 6.
[90] William Sturtevant, "First Visual Images of Native America," in Fredi Chiappelli, Michael J. B. Allen, and Robert Louis Benson (eds.), *First Images of America: The Impact of the New World on the Old*, 2 Vols (Berkeley and Los Angeles, 1976), I: 417–454, 417–419.

in widely disseminated prints by the likes of Holbein, Dürer, and Burgkmair.[91] One could also point to a vibrant local tradition associated with triumph, as for example the roughly contemporaneous sculptural program of Petrarch's triumphs decorating the interior courtyard of Rouen's Hôtel de Bourgtheroulde.[92]

With regard to the particular scenes represented, sources are similarly mixed – ranging from possible first-hand experience (e.g. the naked Tupinamba in the middle and toward the end of the frieze), to reliance on visual clichés, commonplaces, and often-used anecdotes in well-known prints. The Tupinamba at the beginning of the frieze, for example, are not dressed like Tupinamba, but like popular images of them. Though coastal peoples from the areas where Norman merchants traded wore feather headdresses, and sometimes cloaks, they did not wear feather skirts. The feather skirt was, rather, almost ubiquitous among early depictions of the savage, which combined and often melded expectations with information culled from second and third-hand observation. Ceremonial feather scepters such as the one carried by the Tupi in the frieze existed, for example, amongst the Munduruku of the Amazon river basin, but it is unlikely that its inclusion has any basis in a direct – eyewitness – observation. Rather, I think, it can be traced to the widely disseminated 1505 Augsburg edition of Vespucci's *Mundus Novus*,[93] or perhaps, the image accompanying the Antwerp edition of the same book by Jan van Doesborch in 1511 (see Figures 1.4 and 1.5). Both these woodblocks are far from veridical – natives are shown with feather skirts, beards, and European bows, and cannibalism is depicted as a casual, perhaps quotidian, activity, rather than a relatively rare ritual consumption of enemies; nevertheless, both capture a certain residue of first-hand experience, e.g. the feather headdresses, the neck ornaments and the lower arm feather bracelets and anklets that more or less accurately represent Tupi attire. Another detail, one unique to these woodcuts, is the feather tipped scepter that seems to have been directly transposed into stone by the sculptor of Saint-Jacques' frieze (though the Dieppe

[91] On the influence of Bordone's woodcuts on Burgkmair's frieze see Leitch, *Mapping Ethnography*, 82–85.

[92] For more on triumphal themes in Norman (particularly in Rouen) art (windows, bas-relief, manuscript miniatures), and more generally, in civic culture, see Wintroub, *A Savage Mirror*, esp. chapter 4.

[93] See, for example, Feest, The People of Calicut, 297–298; Sturtevant, "First Visual Images," and Rudolph Schuller, "The Oldest Known Illustration of South American Indians," *Journal de la Société des Américanistes* 16:1 (1924): 111–118.

Figure 1.4 Jan van Doesborch, Woodcut broadsheet, *De novo mondo*
(Antwerp, originally 1511, here 1520).

scepter also looks curiously like a *fleur-de-lis*, which was a commonplace image of French kingship).

Another detail, the palm in the Tupi's left hand, seems to be one of the
sculptor's own embellishments. The palm is probably far removed from
any direct experience of the Tupinamba (whether in France antarctique
or transported to Normandy), as it was a commonplace symbol associated with triumph that could have been borrowed from any number
of images (or texts) circulating in Normandy in the 1520s, including
windows, miniatures, and prints, as well as civic rituals, and even poems,
such as that read by the "joueur d'instruments, trompette, crieur public à Rouen," Florent Coppin before the Puy de palinod in Rouen in
the 1520s or early 1530s: "*Le forte palme en triumphe exaltee.*"[94] Just as

[94] Abbé Ch. Guéry, *Palinods ou puys de poésie en Normandie* (Evreux, 1916), 38; for the
poem see Denis Hüe, *Petite anthologie palinodique: 1486–1550* (Paris, 2002), 101–103.
See also Chapter 6. On other sources for triumphal imagery see Wintroub, *A Savage
Mirror*, chapter 4.

Figure 1.5 Anonymous Broadsheet, *Dise figur anzaigt uns das Folck und Insel die gefunden ist durch den christenlichen Kunig zu Portigal oder con seinen Underthonen [This Figure Shows Us the People and Island Discovered by the Christian King of Portugal or His Subjects]*. (Augsburg, c. 1503).

important as skirts and scepters, the contrapposto position of the Tupi holding the scepter is clearly reliant on the well-known artistic vocabulary developed in the popular genre of prints depicting landsknecht and reisläufer. Thus, for example, compare the Tupinamba at the beginning and again in the middle of the frieze with that of a standing foot soldier by an anonymous Swabian Master (c. 1480–90).

Interestingly, many early sixteenth century landsknecht and reisläufer were portrayed wearing elaborate feather headdresses, including images, for example, by Dürer, Leonhard, Burgkmair and Graf. In this sense, they share a certain loose affinity with the Tupinamba of the frieze, but also to Wildmen, such as the one depicted in the triumph carved into the interior courtyard of Rouen's Hôtel de Bourgtheroulde in the 1520s. Wildmen, of course, were among the first and most persistent models for early representations of New World natives.[95] As Stephanie Leitch

---

[95] On the "Wild Man" see, for example, Richard Bernheimer, *Wild Men in the Middle Ages: A Study in Art, Sentiment, and Demonology* (Cambridge, MA, 1952); Susi Colin, "The Wild Man and the Indian in Early 16th Century Book Illustration," in C. Feest (ed.), *Indians and Europe: An Interdisciplinary Collection of Essays* (Aachen, 1987), 5–36.

Figure 1.6 Hans Mair von Landshut, Drawing in ink, "Ein Geharnischter (Oberschwäbischer Landsknecht), um 1500."

has shown, they were also integrally involved with the articulation of a unique German national identity through the representation of Charlemagne as a kind of Wildman or Hercules.[96] This clearly extended to the style adopted by landsknecht, and to "Summer" as he is represented in Rouen's bas-relief of Pétrarch's "Triumph of Time."

One can move beyond stylistic similarities in posturing with weapons, scepters and trees, to place more firmly the frieze in the context of German-Flemish prints by following the image of the Tupinamba with a feathered scepter in Dieppe to the broadsheet *De novo mondo* produced by Doesborch in 1520, a kind of mash-up of Vespucci's text and Balthasar

---

[96] Leitch, *Mapping Ethnography*, chapter 3.

Figure 1.7 Detail, "Frise des Sauvages," l'église Saint-Jacques, Dieppe (c. 1525–1535).

Figure 1.8 Detail, "Frise des Sauvages," l'église Saint-Jacques, Dieppe (c. 1525–1535).

Sprenger's *Die merfart*. The same image used by Doesborch in 1511 (see Figure 1.4) is reproduced in the margins of the broadsheet; more importantly, its other marginal illustrations (see Figure 1.14) consist in relatively crude woodcut copies of either Hans Burgkmair's 1508 woodblock frieze, or Glockendon's knockoff (see Figure 1.13), printed in 1509 and 1511; one or more of these was surely the source that was translated into stone in Dieppe.[97] Neither Burgkmair or Glockendon's contributions to *Die merfart*, however, include images of Tupinamba,

---

[97] J. M. Massing was the first, to my knowledge, to have identified Burgkmair's frieze as a source for the Dieppe sculpture; see his article, "Hans Burgkmair's Depiction of Native Africans," *RES: Anthropology and Aesthetics* 27 (1995): 39–51.

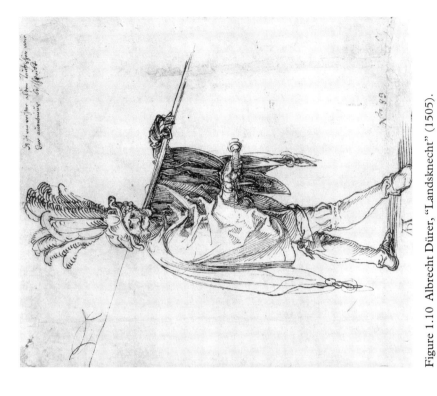

Figure 1.10 Albrecht Dürer, "Landsknecht" (1505).

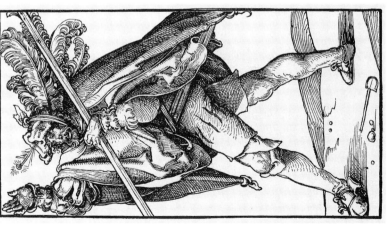

Figure 1.9 Hans Leonhard Schäufelein, "Landsknecht mit geschulterter Helmbarte" (c. 1507).

Figure 1.12 "Summer" from Petrarch's triumphs, sculpture in the Galerie d'Aumale of the Hôtel Bourgtheroulde, Rouen (c. 1523).

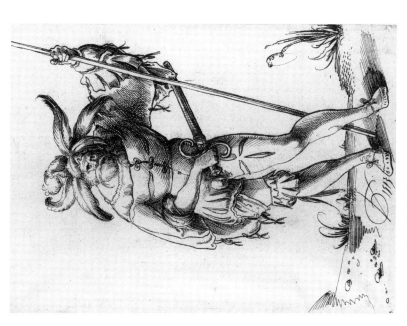

Figure 1.11 Urs Graf, "Reisläufer von vorn gesehen" (1513).

Figure 1.13 Georg Glockendon, "Natives of Guinea and Algoa, natives of Arabia and India."

Figure 1.13 (*cont.*)

rather they begin with representations of "Gennea"; this adds weight to the hypothesis that the source of the Tupi feathered scepter – as well as the other exotic peoples represented in the Dieppe frieze – was Doesborch's broadsheet. Perhaps the presence of a palm frond held by the Indian in the broadsheet, a detail absent from Burgkmair's frieze, but present in the hands of the Tupi in the Dieppe frieze, also speaks to Doesborch's influence.

It is safe to assume, that in addition to Doesborch, Burgkmair, Glockendon and Traut, the author of the Dieppe frieze had access to a wide range of printed sources. One could cite, for example, his decision to include the serpent and the owl in the scene of Gennea, which entwined commonplace biblical images of the fall, such as that of Adam, Eve and the Serpent carved into the western side of church for Ango's private chapel, with an exotic scene of Africa. The sculptor of the Dieppe frieze skips "Allago" (Southern Africa) – which comes next in the Doesborch broadsheet – to carve a relatively close copy of "Arabia." He then borrows ideas, e.g. an Indian bearing a shield and another holding a palm, and translates them into the frieze, before moving on to depict the triumph of the Cochin king.

Preceding the shield bearers and trumpet player who form the rearguard escorting the triumph, loosely modeled after Burgkmair's original or Doesborch's broadsheet, is a man being led in chains. The Dieppe sculptor, apparently knowledgeable about how triumphs were to be represented, placed this captive immediately behind the triumphal train, a detail not included in any of the woodblock prints of the Cochin king, but one which could have been witnessed in any number of civic ceremonial triumphs, or images of them, whether the triumphs produced by Mantegna, or those made by Burgkmair at the behest of the Emperor Maximilian.

Figure 1.14 Jan van Doesborch, *De novo mondo* (1520).

The triumph presented in the Dieppe frieze, though vaguely similar to that in the Merfart cycle, carries not a king on a litter, but a child. There were any number of images circulating around the time that the Dieppe frieze was sculpted that may have influenced the design of this scene and this particular alteration; most notably, an anonymous manuscript of Petrarch's triumphs produced in Rouen's scriptorium around 1503 which included images of a (Christ) child with the Virgin carrying a palm, the "Triumph of Chastity," and a cupid-Christ in the "Triumph of Love."[98] The former would have especially appealed to poets like Crignon and Parmentier who dedicated their verses presented at the Puy de palinod and the Puy de l'Assomption to the Virgin Mary. Another possible referent was the "Triumph of Love" by the Flemish artist, Godefroy Le Batave (see Figure 1.15), who from 1516–26 was closely associated with the court of François I.[99] Recall that Ango was a confidant of the King's sister, Marguerite, and that the King himself visited his quayside home, La Pensée, in 1534. Francesco Collonna's hugely popular *Hypnerotomachia Poliphili*, which included two woodcuts of the Triumph of Cupid attributed to Benedetto Bordone might also have influenced the anonymous sculptor of the frieze.

Bordone was surely familiar to Ango and Co., not only through his well-known woodcuts of Mantegna's *Triumphs of Caesar*, but also – indeed, especially – through his *Isolario* (*The Book of Islands*) printed in Venice in 1528, which included a sketched map of Taprobana, something that would surely have attracted Ango's curiosity. Another candidate, but pushing the limits of the frieze's timeline, might have been the Master of the Die's "Triumph of Cupid" (c. 1530–60).[100] However, I think the most likely source for the anonymous sculptor working in Dieppe was Hans Holbein's frontispiece for Erasmus's *Querela Pacis* (Basil, 1517) (see Figure 1.16). Indeed Leitch believes that Holbein directly relied upon Burgkmair's triumph of the king of Cochin as his model for his triumph of putti. While there is no firm evidence for this, the shared motif of a litter perhaps bears out her supposition.[101] It is certainly among the most likely visual referents in print for the

[98] Bibliothèque nationale de France, Ms. Fr. 594; the illuminations in the manuscript were translated into thread as the Flemish tapestries (c. 1500–23) now held in the Royal Collection at Hampton Court Palace (RCIN 1270). There were, of course, any number of antecedents to these; see Wintroub, *A Savage Mirror*, 214 n.23.

[99] See Myra Orth, "The Triumphs of Petrarch illuminated by Godefroy Le Batve," *Gazette des beaux-arts* (1984), 197–206.

[100] Similarities, confluences and oppositions characterize the relationship between Cupid and the Christ child; to an extent, one comes to stand for the other, with the triumph of cupid, or of love, becoming analogous with the triumph of Christ.

[101] Leitch, *Mapping Ethnography*, 154.

Figure 1.15 Godefroy Le Batave, "Petrarch's Triumph of Chastity," Rouen (c. 1503).

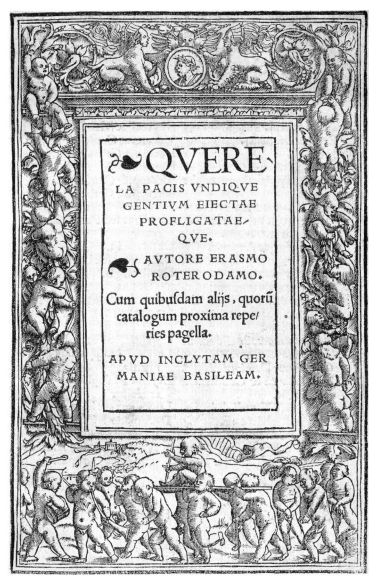

Figure 1.16 Hans Holbein, title-page, Desiderius Erasmus, *Querela Pacis* (Basel: Johann Froben, 1517).

Dieppe bas-relief, which seems to improvise using both Burgkmair's (or Doesborch's) "Der Kunig von Gutzin" (for instance the king of Cochin's headdress, which is transferred from the king to the rear litter bearers) and Holbein's reworking of its themes using putti marching in triumph in place of South Indians.

However, there is also another plausible source for the transposition of the Cochin king into the triumph of a child, which perhaps supplements this reliance on printed images; this can be better approached by following in the footsteps of the frieze's narrative.[102]

The Presence of the Word: Ango Peregrino

The *Frise des sauvages* presents, I believe, a visual representation of Christian translation – a militant theology of expansionist power. It includes not just scenes of natives, families and rituals – Tupinamba, western and southern Africans, East Indians marching in triumph, trees, birds, and monkeys – flowing like a wave across the wall above the trésor, but also a Frenchman, a giant of a Norman (in comparison to the natives before him), leaning into its movement from outside (see Figure 1.17). His body and head are positioned on the curve of an engaged spiral column, following its motion and twisting into the wall and then into the relief itself. Carried over by the momentum of the pillar's winding arc, the man enters the frieze's forward motion; he appears to be on his knees, as if in prayer, one arm outstretched to touch the diminutive (by comparison) Tupi's shoulder, while his other hand clutches a large book. In addition to the book (recalling Saint-Jacques's apostolic mission to Galicia), one can also see the strap of a traveller's bag hanging across his shoulder. Could this be Jean Ango, or might it be Saint-Jacques, or indeed, Jean Ango as Saint-Jacques:

---

[102] The frieze does not stand alone, but crowns stone sculpted into intricate designs, flamboyant tracery, complex foliations – *soufflets*, *mouchettes*, *arabesques*, ribbons and vines, *putti* and figures damaged beyond recognition: heads lopped off by Protestants protesting, rioting, years later (c. 1563) against Catholic idolatry. This, in turn, is set off, framed from below, by eight huge *coquille St. Jacques*. The sculptural program is so damaged that it is difficult to read with any coherence; many of the figures seem exotic, somewhere between human, bear, child, and ape; others are utterly destroyed. Yet each of the larger figures built into the wall is perched on a smaller coquille St. Jacques. Given the ubiquity of the symbol of the Saint and his pilgrimage it seems reasonable to assume that the sculptural program concerns Saint-Jacques' role as an apostolic pilgrim-conqueror; thus, perhaps, setting the stage as it were for the "frise des sauvages" above it as a narrative about pilgrimage, (con)quest, and conversion.

Figure 1.17 Detail, "Frise des Sauvages," l'église Saint-Jacques, Dieppe
(c. 1525-1535).

Ango *peregrino,* bringing the Word to all the world's peoples? Indeed,
his gesture reaching out to the Tupi is accompanied by what appears
to be a confidence being uttered; he's saying something to the native,
grasping his shoulder to ensure his attention. The frieze thus begins
with a depiction of the Word as it travels from a Tupi warrior standing

at its beginning to an African Adam and Eve. The African Adam here, rather than succumbing to evil, is depicted in the act of killing or threatening to kill either a serpent or an owl. Serpents and owls both shared associations with the devil, the serpent less ambiguously so, as owls also symbolized wisdom, and were considered, as George Ferguson suggests, an "attribute of Christ, Who sacrificed Himself to save mankind," thus explaining, he argues, "the presence of the owl in scenes of the Crucifixion."[103] The owl, in any case, appears as if it is about to flee into the next scene, to another family – an Arabian one – thus continuing the flow of information on, through a chain of relations, to a procession marching across the wall carrying a child-king in triumph on a litter, as if he were a relic, a king, or the presence of the Christ child himself. In this reading, and indeed, in the elaborate theology informing medieval and Renaissance triumphs, the triumphator was the instantiation of the word of God – quite literally the Word incarnate.[104] There was, moreover, a close entanglement among various "triumphal-type" ceremonies linked to adventus, translation, and arrivals in state.[105] Presence was denoted situationally – by incorporation into these ritual triumphs – whether a king's triumphal entry, the carrying of the effigy in a royal funeral ceremony, a relic's triumphant translation or that of the Eucharist in *Corpus Christi* celebrations (see Figures 1.18 and 1.19).[106] This is perhaps why Jean Parmentier dressed "in triumph" when he first stepped ashore in Sumatra (see Chapter 5),[107] and why he made such a point of meeting with Ticou's archpriest to "explain" to him that "God had sent his divine Word (*son Verbe*) making himself flesh on earth as incarnated by a Virgin through the operations of the Holy Spirit."[108] Parmentier clearly saw himself as the bringer of a message – a carrier of the Word – to the world; a similar role, I believe, was being played by the child being born triumphally across Saint-Jacques' *nef*. The kingly child in

---

[103] George Ferguson, *Signs and Symbols in Christian Art* (Oxford, 1955), 22.

[104] See Wintroub, *A Savage Mirror*, 102–113.

[105] See Ibid., chapter 7.

[106] Ibid., see also Peter Brown, *Society and the Holy*, 183, and Miri Rubin, *Corpus Christi: The Eucharist in Late Medieval Culture* (Cambridge, 1991), 244. In the period under discussion, with the spread of reformed thinking about the nature of presence, the Church doubled down, following the lead of popular practice; thus when the Council of Trent suggested in the middle of the sixteenth century that triumphal processions be held for the Eucharist, it was articulating as official practice what was already being done in the streets by believers.

[107] See Wintroub, *A Savage Mirror*, chapter 7.

[108] Crignon, *Oeuvres*, 44. See Chapter 5.

Figure 1.18 Louis IX and the translation of the Crown of Thorns to Sainte Chapelle. *Les Riches heures de Jeanne de Navarre* (c. 1336-1340).

the frise des sauvages can thus be understood in Eucharistic terms as the instantiation of Christ's real presence – the Verb. Indeed, the host and the Christ child were closely intertwined, one becoming the other at the moment of transubstantiation, e.g. with the elevation of the Host during the Mass or at the hands of Jews desecrating it.[109]

---

[109] Rubin, *Corpus Christi*, 136–139; and Wintroub, "Translations: Words, Things, Going-Native and Staying True," *American Historical Review* 120:4 (October 2015): 1185–1217.

Figure 1.19 Arrival of emperor Charles IV in Paris, 1378. Grand Chroniques de France (c. 1375–1380).

One can presume a similar ritual instantiation of presence in other ceremonies of the adventus-triumph type, such as that depicted by the frieze. The words spoken to the Tupi at the beginning of the frieze are thus incarnated – and celebrated – by the child-king's triumph at the frieze's end. Propagated across a procession of exotic peoples as if they were distinct, but interconnected, nodes in the flow of information, the Word's signal thus travels and ultimately takes shape materially in the child carried in triumph as would the baby Jesus in the Eucharist, or a Saint in her relic.

The Word's transmission, however, hasn't yet reached beyond the triumphal procession, to the frieze's final figures: primitive looking, or at least sculptural representations that appear – by comparison to the others before – to be rough and unfinished, perhaps deliberately so.[110] From the child-king's triumph, the wave of the frieze thus rolls up against the end of the wall, into men, who are not (yet?) men, who are hardly human, that Vitet describes as "*êtres bâtards et disgracieux*," beyond even the possibility of redemption – or who have yet to hear Christ's word – hence their despair. They are completely naked, clearly unhappy, with the last figure turned to the wall, face in hands as if weeping in shame or sadness; the scene – and the mood – they convey is reminiscent of Masaccio's expulsion of Adam and Eve from the garden of Eden found in Florence's Brancacci Chapel (c. 1425) (see Figures 1.20 and 1.21).[111]

Archives and Savages

Just as the Frise des sauvages followed the path of the Word's instantiation across the world, the trésor that it decorated was a site that embraced and enclosed the presence of the Word in the church. Typically housing relics, the sacred texts that pertained to them, charters, account books, titles, cartularies, and other precious texts and objects of great value and symbolic significance, a church's trésor was a storehouse, a repository, a proto-archive where information was collected and stored and where sacred things were kept.[112] Figure 1.18,

---

[110] In contrast to what Vitet argues, *Histoire de Dieppe,* 266–267.

[111] Several commentators have described these last figures as apes or monkeys; this is possible, but compared with the earlier appearance of a monkey in the frieze, this seems less likely.

[112] Etymologically rooted in the Greek – *thesis*, and the Latin, *thesaurus*. The trésor has been closely associated with the development of municipal, royal and ultimately state archives, e.g. the Trésor des Chartes. See, for example, Olivier Guyotjeannin et Yann Potin, "La fabrique de la perpétuité le Trésor des Chartes et les archives du royaume

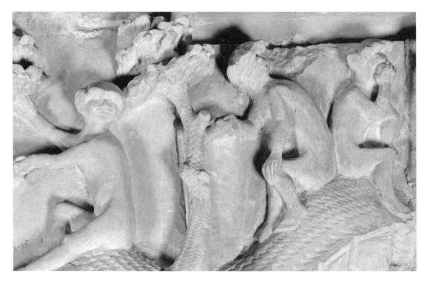

Figure 1.20 Detail, "Frise des Sauvages," l'église Saint-Jacques, Dieppe (Dieppe, c. 1525–1535).

for example, depicts Louis IX's translation of Christ's Crown of Thorns to the Sainte-Chapelle, where not only France's most precious relics were deposited, but where written records having to do with the royal administration of lands, titles, taxes, and privileges were stored. The physical proximity of relics, sacred texts, and administrative records, points to the symbolic entanglement of earthly and spiritual power, the numinous and the word, and to the historical identification of Sainte-Chapelle's collections as the trésor des chartes, which was said to be "*super omnes thesaurus rerum temporalium*" (above all temporal treasures).[113] In the case of Dieppe, David Asseline (d. 1703) refers to the trésor as "the place where the deliberations having to do with the business of the parish were held."[114] Elsewhere, the trésor is referred to as

---

(xiiie-xixe siècle)," *Revue de synthèse* (2004): 15–44; and also Krzysztof Pomian, "The Archives: From the *Trésor des chartes* to the CARAN," in Pierre Nora (ed.), *Rethinking France: Les Lieux De Mémoire*, Volume 4: *Histories And Memories* (Chicago, 2001), 27–100.

[113] Guyotjeannin and Potin, "La fabrique de la perpétuité," 24; as they put it: "La rhétorique politique qui consiste, depuis les années 1330 au moins, à assimiler dans les termes le fonds de chartes au trésor de reliques et de joyaux de la Sainte-Chapelle, trouve ici un point maximal d'expression, puisque les archives du roi sont décrétées *super omnes thesaurus rerum temporalium*, 'au-dessus de tous le trésors temporels.'" Ibid., 31.

[114] Asseline, *Les antiquitez*, I: 115.

Figure 1.21 Detail of Masaccio's expulsion of Adam and Eve, fresco, Brancacci Chapel in the church of Santa Maria del Carmine (Florence, c. 1425).

the sacristy, thus alluding to its role in holding the Church's relics and sacred objects, as well as its parish records. The frieze, then, can perhaps be seen as an inscription that marked a site of reference, that is, a site where presence was enclosed and protected, but also mobilized to make manifest – in information's double movement from collection to deployment – as sacred and this-worldly authority.

Though the frieze's collection of human types – including their weapons, their forms of dress, their family life, their rituals and their distinctive physiognomies – might invite the modern spectator to see it as an incipient form of ethnology, this is an invitation that should be treated with a great deal of circumspection. The information about exotic peoples carved into Saint-Jacques' trésor was not (primarily) about imparting either an understanding of human cultural and geographic diversity or its relativistic consequences. Rather, the story being told employed the exoticism of peoples from Africa, the Indes, and America the same way as it might have used monsters, marvels, and wonders – as compelling inducements and provocations to follow a spiritual (and, in this case, an imperial) narrative: a quest to imagine, materialize and extend God's Word throughout the world.[115] Like the *trésor* that it decorated, it signaled the entanglement of sacred and worldly power in the collection, propagation, and extension of the Word. At the same time, the frieze was also recounting a social journey – the translation of its *mécène*, Jean Ango, into the earthly counterpart of Saint-Jacques, the imperial pilgrim.

This being said, the serial quality of the frieze flowing across the wall certainly suggests a quasi-systematic attempt to record, classify and collect information brought back by men sailing for Ango. In this sense, one might see the frieze as a kind of distant cousin of modern ethnological practice insofar as it served as a kind of spiritualized analogue of activities of merchants and explorers in collecting intelligence about potential trading partners, goods, and faraway places. Accordingly, the frieze acted as a conduit between the particularities of the empirical world (*historia*) collected by the likes of Jean Parmentier and Pierre Crignon, and *scientia* – true knowledge of universals as enshrined by the "Queen of the Sciences," Theology.[116]

Dieppe's *frise des sauvages* was a work of mediation, of translation, that coalesced, integrated and extended information originating (supposedly) in the experience of firsthand observers. Burgkmair, for example, likely met with Balthasar Sprenger and with others that participated on his voyage. Moreover, he (like Dürer) had had the opportunity to view commodities and artifacts (and perhaps even peoples) brought back from Africa, the Indes, and America.[117] Yet Burgkmair's extraordinarily successful frieze was clearly a work of improvisation that integrated, transformed and inflected this information – by comparison, selective appropriation, and synergistic analogy – with the conceptual schemas and artistic traditions operative in his intellectual, cultural and social milieu(s).[118] Similarly, while connected to

---

[115] See, for example, Sandra Young's "Envisioning the Peoples of 'New' Worlds: Early Modern Woodcut Images and the Inscription of Human Difference," *English Studies in Africa* 57:1 (2014): 33–54.

[116] This distinction, and its mediation, is the subject of Chapter 2.

[117] Note, for example, Dürer's well-known reactions to seeing Aztec treasures sent by Cortés to Charles V. See Jean Michel Massing, "Early European Images of America: the Ethnographic Approach," in Jay A. Levenson (ed.), *Circa 1492: Art in the Age of Exploration* (New Haven, 1991), 515. On Burgkmair's direct experience of indigenous peoples and their artifacts, see Leitch, *Mapping Ethnography*.

[118] See, for example, Philippe Descola, *Beyond Nature and Culture*, trans. Janet Lloyd (Chicago 2013). John Coltrane meets Pierre Bourdieu. To paraphrase the cognitive anthropologist, David White, "jazz is spontaneous expression; it is not *unfettered*

Figure 1.22 Graffiti Ship on the trésor of l'église Saint-Jacques (early sixteenth century, Dieppe).

spontaneous expression (contrary to folk belief), because it is always constrained by form, meter, or sonic context." On experience and tradition in Burgkmair's frieze see, for example, Peter Mason, *The Lives of Images* (London, 2001), 89ff; Ashley West, "Global Encounters: Conventions and Invention in Hans Burgkmair's Images of Africa, India, and the New World," in Jaynie Anderson (ed.), *Crossing Cultures: Conflict, Migration, Convergence*. Proceedings of the 32nd Congress of the International Committee of the History of Art (Melbourne, 2009), 272–279, and Massing, "Hans Burgkmair's Depiction."

voyagers and their manifold experiences in faraway ports, the Dieppe frieze was also (very) far removed from them. In the nineteenth century Ludovic Vitet registered his surprise that such a profane subject – with no miter, priest or cross anywhere to be seen – could have found its way onto the wall of a church;[119] but, what he sees as an entirely secular monument to the maritime exploits of Jean Ango and the Dieppois was, in fact, a deeply religious representation of Christian imperialism. Far from being simply a proto-type of an ethnographic collection, the exotic peoples found marching across Saint-Jacques' nef presented sixteenth-century observers with a reason to voyage across the earth and a reason to hold up the frieze's mécène (Jean Ango) as a doer of God's work. On the one hand, its purpose was to provoke, enlist, inspire and replicate the apostolic extension of the Word throughout the world. On the other, it was to valorize and legitimize the commercial exploits of Ango and his men. The frieze was thus an embrace of Saint-Jacques' imperial-theological quest; however, unlike the conquest represented by images of Santiago Matamoros who bloodily triumphed over Muslims, trampling them beneath his horses' hooves and carving them up with his sword, the Saint-Jacques envisioned here was the seemingly more pacific – and certainly, more bookish – Santiago peregrino, owing as much to the content of Erasmus's *Querela Pacis* as to Holbein's woodcut frontispiece.

Ships of Stone

Below the frieze, just a few feet off the ground, another more literal representation of a ship is to be found etched into the wall of Saint-Jacques' trésor: a late fifteenth, early sixteenth-century carrack, similar, perhaps, to La Pensée. Unlike the frise des sauvages, however, this graffiti ship was not made by an artisan – an "expert" in sculpting stone – but by a man of the sea on his knees with a dull knife. The ship was scratched with a surprising degree of precision, both in terms of scale and detail. It has three masts (square sails on the fore and main masts, both with top sails, and a lateen rigged mizzenmast), at the bow, there is a place for a lateen spritsail to be attached to a jib boom under the bowsprit; this is matched at the aft with a small lateen sail behind the mizzen. Standing rigging – stays and shrouds – supporting

---

[119] Vitet, *Histoire de Dieppe*, 261.

the masts, top masts, and yards is shown (e.g. the mainmast has five shrouds and two stays), as is the running rigging (sheets, halyards, topping lifts). It has two large castles and six canons (presuming a symmetrical emplacement on the other side). The square sails appear to be furled, though the aft mizzen seems to be deployed; this is something that would probably only occur while maneuvering close to land, a supposition that is perhaps borne out by the tautness of the line attaching a small boat carrying three barely discernable men to the ship from behind, though it would be injudicious to read too much into this sketch in stone. Nevertheless, unlike the anonymous author of the frieze, it is unlikely that Saint-Jacques' graffiti artist drew inspiration from prints, books and civic ceremonial, to craft his design. This is an image constructed by a well-trained eye habituated by experience to sailing ships over long distances.

It is impossible to say which came first – the frieze or the graffiti. Nevertheless, each in their own way depicts a voyage as a spiritual offering: a prayer carved in stone. The frieze, as we have seen, was a kind of map and a guide – a meditative conduit and contemplative pilgrimage where those below could touch with their eyes distances only imaginable in an abstract space of representation. The scaled graffiti ship encapsulated a similar kind of voyage – sailing both on a sea of stone and on the open seas: an *ex voto* offering, a sacred talisman, an amulet, designed to protect or commemorate intrepid sailors on their journeys to distant lands. In this regard, both frieze and ship had much in common with the poetry read by Jean Parmentier and Pierre Crignon for Dieppe's Puy de l'Assomption, which met each year on August 15 only steps away in Saint-Jacques' nef.

The Poetry of Ships

Like the "frise des sauvages" carved high up on Saint-Jacques nef, and the graffiti ship carved at knee-level below, the poetry Parmentier and Crignon read out loud before an audience of fellow poets evinced both the experience and expertise of men who knew a great deal about ships and how to sail them, and the integration of this knowledge with the universal truths of Christianity. Take, for example, Parmentier's poem, "The great route to the port of true salvation" where he recounts how "a great and well-equipped ship" was caught (North-by-North-East) at the evening tide by a terrible, Satanically inspired, storm.[120] The crew of the ship, he tells us, "turned the helm to port," but this only took them further off course. The ship heaved and rolled; the wind tipped her this way and that; they were on the verge of losing all, when the quick-thinking crew, "each doing his part," maneuvered "cables and thickly braided ropes (greslins)" and "made ready the anchor" to use as a counterweight to "tow" the ship out of danger, and back on course – to "the great route to the port of true salvation." Parmentier goes on to explain in the poem's third stanza that humanity was this ship, which had been set on the wrong course by Eve's sin and by an evil wind emanating from the pride and envy of Satan, the serpent. However, through the patronage of the Virgin, he says, the ship would be put back on course, toward the port of true salvation.

What is interesting, in the present context, is not only how Parmentier inflected his poetic ship as a theological metaphor for translating humanity across the troubled seas of Satanic perfidy, but how he was able to integrate his expertise and knowledge of ships and navigation into verse. Thus, his poem describes kedging (or warping), a difficult procedure of fixing a tow-line to an anchor, often set by a small boat, that was then used to pull a ship against the wind (or in a dead calm) out of dangerous waters. Kedging was hardly a technique that would

---

[120] Parmentier, *Oeuvres*, 37–39.

> Dessus la mer, jadis, faisoit sa route
> une grand nef assez bien esquippée,
> d'un si bon temps, que bolline ou escoute
> n'avoit mestier d'estre prinse ou happée,
> tant qu'il survint, environ la vesprée,
> de l'Est Nord Est, une horrible tempeste,
> don't le patron eust bien mal à sa teste.
> son mathelot mist heaulme à bas bort,
> mais ne fut pas qinsy qu'il le faillut,
> car par cella laisserent à thiebort
> le grand scenail du vray port de salut.
>
> La tempeste dedens la nef se boute
> tant qu'elle fut sur les bancz eschouée
> et le grand vent la boute et la deboute,
> don't mathelotz, craignantz d'estre trouée,
> legierement feirent une touée,
> cables, greslins, on auste et apreste,
> à l'esquelt bien et l'ancre toute preste,
> at à haller chascun feit son effort,
> par quoy cela sit res bien leur vallut
> qu'ils congneurent, par leur maneuver fort,
> le grand scenail du vray port de salut.

have been widely known to those who were not experienced sailors. The same could be said for the detailed representation of the carrack found scratched on Saint-Jacques' nef, which intimates an expert's knowledge of masts, sails, and spars and how they were to be maintained, managed, and maneuvered by dozens of men and miles of rope.[121] It was this kind of practical information and knowhow that made voyages possible.

Voyages, however, were construed not only in geographical and commercial terms, but also as spiritual journeys – as pilgrimages.[122] Surely, it was this schema of quest, conquest, pilgrimage, and hunt that formed, informed and inspired Parmentier's voyage to Sumatra. Nefs within a nef, the frieze and the ship scratched in the wall below it, like Parmentier's devotional poetry, were all inscriptions that acted to transport Saint-Jacques' voyagers across oceans and worlds, just as the trees in the *Art of War* transported Machiavelli and his companions back to ancient Rome.

At the same time as they referred to spiritual-geographical journeys, the frieze, the ship, and (as we will see in greater detail in the next chapter) Marian verse, inscribed, enacted, and mobilized material, epistemic, and social voyages. In this sense, they worked together to translate the disparate facts and sublunar particularities of the terrestrial world (as "discovered" by eyewitnesses, or found in prints and texts) into universal truths by embedding them simultaneously in both the physical architecture of Dieppe's most important Church and in theological-geographical narratives about the extension of God's Word throughout the world. In a similar fashion, they also mapped out voyages that transgressed accepted notions of rank and status to carry ambitious merchants, such as Jean Ango, across and up social hierarchies of power, while also legitimating, as a spiritual quest, their commercial and

---

[121] Failure to manage the complex array of environmental and shipboard factors could be fatal. This was complicated hands-on knowledge that was neither written about nor codified in early sixteenth-century France, rather it was imparted sailor to sailor in the field and on the job. For example, with the wind behind it, the square sail on the foremast, placed in front of the ship's center of gravity, had a tendency to cause the ship to pitch, heave, and/or be carried off wind; however, if balanced against the work being done by the mizzen, placed behind the ship's center of gravity (mizzen means balance in Arabic), the ship's forward motion could be stabilized. In other words, a ship of the complexity of the carrack scratched into Saint-Jacques' wall required an expert coordination of wind, bodies, cords, spars, and sails (see Chapter 5).

[122] For example, to "The Perfect Port of Salvation and Joy" (ibid., 27); to the "*Terre Neufve* in All Ways Fruitful" (ibid., 40); and to the "New World, Always Pure and Holy" (ibid., 46).

cultural activities. Signaling not only the empirical presence of extended networks of transoceanic navigation, the materialized prayers located on – and read aloud in – the nef of Dieppe's l'église Saint-Jacques (or in Rouen's Cathedral) can thus be understood (like the astrolabe that is the subject of the chapter that follows) as instruments of imperialism and empire, that is, of information.

2 Expertise: The Heavens Inscribed

What is an expert? Though much recent work has been devoted to this question, I have in mind a relatively little-studied dimension of the expert's expertise.[1] Derived from the Latin *expertus*, "expertise" is generally defined as the possession of specialized knowledge and skill. This is well known. However, if we travel back to the early modern period, the word had several other quite interesting and unexpected associations. For example, in his *Thresor de la langue françoyse* (1606), Jean Nicot defines an expert not only as one who is knowledgeable and/or practiced in many things, but also as someone particularly ingenious (*ingenieux*), which he defines as "one who has a good mind and understanding."[2] He adds further precision to this definition with several synonyms: "*artificiosus, argutus, solers.*"[3] The first word of this trinity takes us toward a less discussed and darker side of expertise. While artificiosus retains the strong connotation of skill and knowledge still associated with expertise today, it was also defined as "*ruse, deguisement, fraude.*"[4] We see this same set of meanings trace itself even further back into the sixteenth century, as illustrated by Robert Estienne's *Dictionarium latinogallicum* (1538), where, like Nicot, Estienne defines the expert both as one "*qui a veu et faict beaucoup d'experiences et essaiz*" and as argutus – that is, one with a subtle and ingenious mind.[5] Estienne then cites a number of ancient authorities to describe argutus (*subtil, ingenieux, agu*) as "*trop affectee, trop diligente et curieuse,*" as "thin" (*maigre*) and as "birds that make a lot of

---

[1] For example, E. Selinger and R. Crease (eds.), *The Philosophy of Expertise* (New York, 2006); M. Lynch, "Circumscribing expertise: membership categories in courtroom testimony," in S. Jasanoff (ed), *States of Knowledge: The Co-production of Science and Social Order* (London and New York, 2004), 161–180; and H. Collins and R. Evans, *Rethinking Expertise* (Chicago, 2008).

[2] See Nicot, *Le Thresor*, s.v. expert.

[3] Ibid.

[4] *Dictionnaire de L'Académie française, 1st Edition* (1694), 58. Thus Sir Richard Barckley qualifies certain men as "artificiall apes" who counterfeit "a formall kinde of strangers civilite." Richard Barckley, *Discourse of the felicitie of man* (London, 1598), 327.

[5] Estienne, *Dictionarium*, s.v. *expertus*.

noise" (*les oiseaulx font grand bruit*).[6] Here, too, experience is inflected with the suspicion that claims of superior knowledge and skill were in reality the mark of an imposter, a maker of fakes, a producer of lies, a parvenu. Thus, to the usual definitions of an expert as one who was "much experienced in things," or who was considered "skillful," were grafted other less flattering associations. As Randle Cotgrave said, an expert is someone "cunning" and/or "well seene."[7] Cunning, like ing-enieux, had positive connotations, of course, but it too shaded into the pejorative with associated meanings such as *regnarder* or *ruser* – that is, someone shifty, crafty or practiced in sleight of hand.[8] "Well seene," like Estienne's "*trop affectee*," adds to "cunning" the desire to appear in a favorable light.[9] To be "well seene" similarly suggests an element of deception and dissimulation, of feigned display and/or trickery, someone who "perverts a truth with shifts, trickes, or subtilties."[10] This implies, once again, that the expert's desire to be "well seene" might also be asso-ciated with purposive attempts to garner power and prestige through guile and perhaps even fraud; in other words, that the "expert" might be considered something of a social climber, as hinted at, not so subtly, by Estienne's characterization of the subtle and ingenious mind as "*fine et affectee*."[11]

There can be little doubt that the negative associations bound up with expertise were closely associated with its relationship to experience. As Cotgrave defined it, experience is "cunning, skill, knowledge, wisedome, gotten by much practise, and many trialls."[12] Experience, here, had less to do with naked acts of perception than with hard work, practice and wisdom gained through careful and repeated rehearsal. This definition of experience is somewhat at odds with our own, though perhaps not entirely opposed to modern notions of expertise. In early modern Europe, experience was understood to refer to how nature usually, i.e. universally, behaves, not to perception, empirical investigation or induction from dis-crete events.[13] Aristotle's *Posterior Analytics* is the touchstone here:

One necessarily perceives an individual at a place and at a time, and it is impos-sible to perceive what is universal and holds in every case. Since demonstrations

---

[6] Ibid., s.v. *argutus.*
[7] See Cotgrave, *A Dictionarie*, s.v. expert.
[8] Ibid.
[9] Ibid.
[10] Ibid.
[11] Estienne, *Dictionarium*, s.v. *argutus.*
[12] Cotgrave, *A Dictionarie*, s.v. *experience*, 419.
[13] P. Dear, *Discipline and Experience: The Mathematical Way in the Scientific Revolution* (Chicago, 1995), 20–21.

are universal, and it is not possible to perceive these, it is evident that it is not possible to understand through perception.

Rather, "from perception there comes memory ... and from memory (when it occurs often in connections with the same thing), experience; for memories that are many in number form a single experience."[14] Thus experience, for Estienne, was synonymous not only with expertise, but with proof (*espreuve*), based on reference to common knowledge rather than on the assertion of a personal experiential claim. Experience of discrete phenomena was not really knowledge at all, for it was thought to be concerned with ephemera, particulars and singularities, or, as they were frequently referred to, as monsters.

According to the OED, the word "monster" derives from the Anglo-Norman and Middle French *monstre*, or *moustre*, for prodigy or marvel, but also for a disfigured, misshapen being or one that is "contrarie to nature."[15] A monster might refer to a marvelous wonder, a prodigy, or a miracle in which the work of God's (or perhaps the Devil's) hand could be discerned; or it could refer to the credulity and rudeness of ignorant and uncivil wonder-seekers; or worse yet, men who were incapable of distinguishing between – or who deliberately confused for their own purposes – spectacles of the monstrous and the presence of the divine. The truth of the monster was exceptional; it wasn't *scientia*, but it was, or could be, a spur to seeking it out. In this sense, the monster might be seen as a transitional figure, one that, like wonder, needed to be tamed so as to mediate between the ephemeral singularities of the material world and universal truths. Monsters were, accordingly, close cousins of information – singularities that were "shapeless, ill-favoured, fashionless; ouglie, rude,"[16] but which could also, like information, refer to proof, to facts, and to intelligence.[17] At the same time, monstre was a performative act of demonstration – the "showing," "expressing," or "representing" of such proofs.[18] The history of the word can thus be traced beyond its

---

[14] Quoted in ibid., 22, from Aristotle, *Posterior Analytics*, in J. Barnes (ed.), *The Complete Works of Aristotle: The Revised Oxford Translation*, 2 vols (Princeton, 1984), I: 31. See also P. Dear, "Mysteries of State, Mysteries of Nature: Authority, Knowledge and Expertise in the Seventeenth Century," in S. Jasanoff (ed.), *States of Knowledge: The Co-production of Science and Social Order* (London and New York, 2004), 206–224, esp. 207–9.

[15] OED, s.v. monster.

[16] See Chapter 1, note 3.

[17] Cotgrave, *A Dictionarie*, s.v. *monstre*. Here too, there are close filiations with information as fact, proof, or an item of data; as in the London merchant adventurer, Robert Thorne's 1527 *Declaration of the Indies*, "being an Information of the Parts of the World, Discouered by Him and the King of Portingall...." or in Nicot's definition: "*informations faites et rapportées,*" *Le Thresor*, s.v. *monstre*.

[18] Cotgrave, *A Dictionarie*, s.v. *monstre*.

Anglo-Norman usage to *monere*, that is, to prodigious facts and beings, the evocation of fear and/or wonder, and the intervention of omens that warn, betray or advise.[19] In this regard, proofs might be either "revealed" or "betrayed." To betray, in the sense of to show or to warn, entangled monstre with a number of pejorative associations having to do with deceit, treachery, and deception. As with information, the ambiguity of the thing, and the act, had the potential to color – validate, legitimize or discredit – the people associated with it, and of course, vice versa. Such translations could easily cross into darker territories: into credulity, idolatry, and self-aggrandizing social promotion. Nicot, for example, defined monstre as "*dressé et fait ou habitué à faire grande monstre de soy.*"[20] A monster, then, might refer not only to contingent experience or to "monstrous" anomalies, but also to those who, contrary to nature, transgress (betray) the boundaries of expected social or professional behavior; that is, to parvenus seeking – through cunning, guile, and expertise – to rise above their given social stations (*argutus*). The balance between proof and credulity, unnatural singularity and legitimacy, was delicate indeed, and applied to humans as well as to their entanglements with the world. As a matter of course, the hunt for social legitimacy was closely tied to efforts to demarcate epistemic and spiritual legitimacy. The generalization of wondrous singularities into singularities *tout court* (i.e. as the raw and discrete data of particulars) that could be taken as signs of – and evidence for – the ordinary course of nature (*scientia*), thus paralleled the transformation of natural history, medicine, astronomy, and navigation into worthwhile and laudable endeavors.[21] The quest here was not simply to bring God's Word to the world (see Chapter 1), but to develop (and socially validate) cunning experts capable of reading the Word of God *in the world*.

Writing in the first quarter of the seventeenth century, Galileo Galilei comments that philosophy could be found written in the book of the universe. But, he continues, this

book cannot be understood unless one first learns to comprehend the language and to read the letters in which it is composed. It is written in the language of mathematics, and its characters are triangles, circles, and other geometric figures,

---

[19] See Asa Simon Mittman, Peter J. Dendle, *The Ashgate Research Companion to Monsters and the Monstrous* (Surrey, 2013), particularly the contributions by Felton, 103–132 and Steel, 257–274; and Jean Céard's *La nature et les prodiges: L'insolite au XVIᵉ siècle* (Geneva, 1996).

[20] Nicot, *Le Thresor*, s.v. *monstre*.

[21] See the seminal work of Zilsel, *The Social Origins of Modern Science*, ed. D. Raven, W. Krohn, and R. Cohen (Dordecht, 2003); and Robert Westman, "The Astronomer's Role in the Sixteenth Century: A Preliminary Study," *History of Science* 18 (1980): 105–147.

without which it is humanly impossible to understand a single word of it: without these, one goes wandering about in a dark labyrinth.[22]

Whereas Galileo's words intimate the unassailable status mathematics was to achieve in the modern world, this was not always the case. In early modern Europe, practical or applied mathematics was considered by many, especially in university-educated circles, a rather lowly – even menial – form of knowledge, considered only slightly superior to that of a skilled tradesman.[23] It was typically viewed in practical "hands-on" terms – e.g. mensuration, surveying, ballistics, astrology, and navigation. Its practitioners inhabited the relatively lower rungs of the intellectual hierarchy. This was so for a variety of reasons. First, true science (*scientia*) was thought to be based on universals, whereas practical mathematics was seen either as a descriptive method applied to contingent earthly phenomena (i.e. singularities/monsters), or as the logical working out of relations between purely imaginary mathematical entities. Put somewhat differently, unlike natural philosophy, mathematics provided no answers to questions of causation.[24] Philosophy alone attained to the heights of true knowledge, though its status too diminished – to that of "hand servant" – when compared to theology, the undisputed queen of all the sciences.

Mathematics, moreover, was also tainted by associations with the occult, and such activities as divination, necromancy, and astrology. As John Aubrey was to comment in the mid sixteenth century: "Astrologer, Mathematician, and Conjurer were accounted the same things."[25] What Katherine Neal has identified as the rhetoric of utility was an important means of disentangling mathematics from these troubling associations;[26] at the same time, in the early years of the sixteenth century, this rhetoric was far from being an effective means for elevating the epistemic status of mathematics or the social status of mathematicians, given that utility, in and of itself, was not seen as particularly valorizing or creditworthy; indeed, it was often understood as being the very opposite – as derogatory. Utility, taken in this sense, was less a resource than a spur

---

[22] Galileo Galilei, *Discoveries and Opinions of Galileo*, translated and edited by S. Drake (New York, 1957), 237–238, from the *Assayer* (1623); Paracelsus made similar observations regarding medicine in the 1530s, see esp. Smith, *The Body of the Artisan*.
[23] See, Westman, "The Astronomer's Role," and M. Biagioli, "The Social Status of Italian Mathematicians, 1450–1600," *History of Science* 27 (1989): 41–95. See also Chapter 4 with reference to Jean Fernel.
[24] See Dear, *Discipline and Experience*, 36.
[25] Quoted in J. P. Zetterberg, "The Mistaking of 'the Mathematics' for Magic," *Sixteenth Century Journal* 11 (1980): 85.
[26] K. Neal, "The Rhetoric of Utility: Avoiding Occult Associations for Mathematics through Profitability and Pleasure," *History of Science* 37 (1999): 151–178.

to elevate the status of practical mathematics.[27] Navigation, for example, though certainly a useful (and potentially profitable) skill, needed to be transformed into a proto-nationalistic, dynastic and eschatological quest before it could become sufficiently laudatory to merit the patronage of kings and princes. Thus, though the "rhetoric of utility" was a crucial condition for the epistemic and social legitimation of practical mathematics, it was not, in and of itself, sufficient. Other means were necessary.

Though expertise in the early modern period was associated with specialized knowledge and skill, it was also associated with cunning, deception and social climbing. Such knowledge potentially threatened well-defined and time-honored social and disciplinary boundaries. This was certainly the case with practical mathematics, which was considered by many to be an inferior grade of knowledge, especially when compared with natural philosophy and theology. This spawned numerous attempts to elevate the status of practical mathematics and to lend legitimacy to its practitioners. This chapter will focus on one such attempt, that by the French cosmographer, explorer and poet, Pierre Crignon. In the pages that follow we will examine how this singular monster, a social climber and parvenu, attempted to rise above his station by transforming his mathematical expertise into creditworthy and laudable knowledge worthy of the respect (and patronage) of powerful merchants, princes and even kings.[28]

A Poetic Astrolabe

Jean Parmentier and Pierre Crignon were well known for their exploits on the open seas; but they were also known for their poetic skills in composing Marian verse. Both, for example, were singled out by their contemporary Pierre du Val, as being among the best poets in all of France.[29] As evinced by contemporary manuscript collections of sixteenth-century poetry, both were frequent competitors at the annual poetic concourses sponsored by Rouen's Puy de palinod and Dieppe's Puy de l'Assomption. Their contributions often won the day. In Rouen, Crignon was crowned laureate in 1517 and 1527 with poems such as

---

[27] On the relationship between imperialism and mathematics see A. Alexander, "The Imperialist Space of Elizabethan Mathematics" *Studies in History and Philosophy of Science* 26:4 (1995): 559–591.

[28] As an interesting point of comparison, see, for example, Robert Westman's seminal article, "Proof, Poetics, and Patronage: Copernicus's Preface to *De revolutionibus*," in D. Lindberg and R. Westman (eds.), *Reappraisals of the Scientific Revolution* (Cambridge, UK, 1990), 167–205.

[29] P. Le Verdier (ed.), *Pierre du Val, Le puy de souverain amour*, originally published 1543 (Rouen, 1920), fol. A iv[v].

Purple, Excellent for Dressing the Great King, and *The King of the Treasures of Eternity*.[30] Parmentier, similarly, was honored at Rouen's Puy in 1517, 1518 and 1528, and at Dieppe's in 1520 and 1527 for poems such as *The Perfect Port of Welcome and Joy*, and *The Strong Ship Completely Filled with Grace*.[31]

Sometime in the first third of the sixteenth century, La Pensée's navigator-cosmographe, Pierre Crignon, read to the assembled members of Rouen's Puy his poem, *Just Astrolabe Where the Sphere Is Comprised*.[32] This chant royal provided a detailed exposition of the many similarities between an astrolabe and the virgin mother of God, comparing every aspect of the astrolabe's design, ornamentation, and use to the Virgin.[33] Both, he explained, were created with reference to the cosmic perfection and virtuous symmetry of the celestial sphere:

> Our astrolabe where the sphere is comprised,
> Is the humble Virgin in her conception,
> The maker is God, who bestows grace upon her,
> For his son to receive.
> The dawn meridian line is her inception,
> The circles round and graduated,
> These are the virtues God confers on her.
> The equator is justification,
> The tropics her glory, exalted
> Of all the virgins [on whom] he bestows honor, he prefers her.
> The zodiac and rete,
> Gifts of beauty and perfection,
> The right angles, the benediction
> Of the All Powerful, who watches over and sustains her,
> Who made her exactly according to his intention
> To contain by his extension,
> The perfect circle that encloses and contains him.
>
> The one who made the great mappemonde
> Well proportioned in zones and climates,
> Watching over human beings on the world's seas
> Disoriented, in desolate peril and affliction,

---

[30] See Crignon, *Oeuvres*.
[31] See Parmentier, *Oeuvres*, respectively 27–29, and 17–20.
[32] Jean Parmentier also wrote a chant royal that concerned the astrolabe – which, he says, made manifest "for the great benefit of all humanity, the high secrets of heavenly motion." See Hüe's *Petite anthologie*, 175–177.
[33] In a similar chant written earlier in the sixteenth century, Nicolle Osment compares the Virgin to a "sphere showing all the secrets of the heavens." See P. Vidoue, *Palinodz, chants royaulx, ballades, rondeaulx, et epigrammes, a lhonneur de l'immaculee Conception de la toute belle mere de dieu Marie Patronne de Normans presentez au puy a Rouen …* (sl, 1525). Reprinted by E. de Robillard de Beaurepaire (Rouen, 1897), fols. xxiv–xxv.

Having lost anchors, sails and masts,
In storms and winds of ingratitude,
Not knowing under which latitude
They were led, navigating astray,
To plot their course following the finest line of gold,
To bring them to a safe and good land,
Just astrolabe where the sphere is comprised.

He turned it and made it round,
Then graduated the limb with his compass.
Of purity he made the face (the mater) profound
To receive tables that are not
Tarnished with sin, for without failing by so much as a pace,
Four right angles he promptly put in their place,
Conferring bountiful beauty upon her.
Each tropic was put in order
And the equator too, without repetition, To confirm her as the king's
    instrument,
Just astrolabe where the sphere is comprised.

On the tympana, turned in gold pure and holy,
Of which beauty surpasses human imagination,
He made with his compass, without error,
The azimuth and the almicanaratz,
The dawn line without illusion,
The right zenith of meek humility,
True horizons, and accurate hours;
Then put beneath the rete carrying
The zodiac's procession of stars,
And the rete's index that shows him the law,
Just astrolabe where the sphere is comprised.

On the other side, it is surrounded with stars,
Where are portraits, by degree high and low,
Signs of love, days and months, where is found
The year in which wars and quarrels end.
Then the alidade is added, with delight,
To demonstrate the phases of the altitude
Of the true sun at its highest point
When it will come, by the line of faith,
To enter into the well sighted pinule.
And then will be found there, as I believe,
Just astrolabe where the sphere is comprised.

From the alchitoth (the pin assembly) where charity abounds
All is conjoined, from which the poor grow weary,
Seeing the sphere in the vast and limitless ocean contained,

they have joy and solace
And they evade death's snares and traps,
Pondering, under such similitude,
That the Virgin and mother, in holy sanctity,
Would contain God and that, by his will
Would be by grace in idea so seized
That one would hardily say, without fear,
Just astrolabe where the sphere is comprised…[34]

---

[34] See Parmentier, *Oeuvres*, 62–65; the poem can also be found in D. Huë, *Petite anthologie palinodique: 1486–1550* (Paris, 2002), 171–174. This poem was not signed, but has been attributed to Parmentier by Ferrand. Denis Huë has since definitively established that it was by Parmentier's navigator, Pierre Crignon. D. Huë, "Un nouveau Manuscrit palinodique, Carpentras, Bibliothèque Inguimbertine nx 385," *Le Moyen Français* 35–36 (1995): 175–230; for a contemporary account of the astrolabe see J. Focard, *Paraphrase de l'astrolabe* (Lyon, 1544).

Nostre astrolabe où la sphere est comprise,
C'est l'humble Vierge en sa conceptïon,
L'ouvrier, c'est Dieu, qui grace lui confere,
Pour de son filz faire receptïon,
La ligne aurore est son inceptïon,
Le cercle rondz et graduacïon,
Ce sont vertus que Dieu en elle infere,
L'equateur est justificatïon,
Tropiques sont gloire, exaltacïon
Dont toute vierge, en honneur, el prefere.
Le zodïac et la retz stillifere,
Dons de beaulté et de perfectïon,
Les angles droiz, la benedcïon
Du Tout-Puissant qui la garde et soustient,
Qui la faict just à son intentïon
Pour contenir, soubz son extentïon,
Le rond parfaict qui l'enclost et contient.

Celuy qui fit la grande mapemonde
Bien compassée en zones et climatz,
Voyant humains en la mer de ce monde
Hors de leur routte, en perilz vains et matz,
Ayant perdu ancres, voilles et matz,
Pour la tempeste et vent d'ingratitude,
Non congoissans soubz quelle latitude
Estoient menéz, navigans en desroy,
Leur compassa du plus fin or qu'on prise,
Pour les conduire en sain et bon terroy,
Just astrolabe où la sphere est comprise.

Il la tourna et fist de forme ronde,
Puis gradua le limbe par compas.
De purité fut la face profonde
Pour recepvoir tables où ne sont pas

A Computer of Brass

The astrolabe, or star-finder, upon which Crignon based his chant, was a model of the heavens inscribed in brass. A kind of analogue computer, it was, at least in theory, capable of facilitating a number of complex astronomical and geometrical calculations.[35] In an era before accurate and portable clocks it could be used to tell time; it could be

34 (cont.)   Traictz maculéz, car sans faillir d'un pas,
Quatre angles droictz y feist en promptitude,
Luy conferant de beaulté plenitude.
Chacun tropique y fut mis en arroy
Et l'equateur justement, sans reprise,
Pour l'approuver comme instrument de roy,
Juste astrolabe où la phere est comprise.

Sur les tympans tournéz d'or pur et munde,
Dont la beaulté surpasse humains caractz,
De son compas fit, sans rasure immonde,
Les azimus et almicantaratz,
La ligne aurore exempte de baratz,
Le droict zenith d'humble mansuetude,
Horisons vrays, heures de rectitude;
Puys mist dessus la retz portant en soy
Le zodïacq d'estoilles entreprise,
Et l'ostenseur qui le monstre, par loy,
Just astrolabe où la phere est comprise.

De l'aultre part, maynte orbe la circonde,
Où sont pourtraictz, par degretz haultz et bas,
Signes d'amours, jours et moys, où se fonde
L'an où prend fin la guerre et toutz desbatz.
Puis l'alidade y a mis, par esbatz,
Pour demonstrer aux quartes d'altitude
Du vray soleil la haulte magnitude
Quand il viendra, par la ligne de foy,
Entrer dendens la pinulle bien prise.
Et lors sera trouvé, comme je croy,
Just astrolabe où la phere est comprise.

De l'alchitoth où charité habonde
Fut tout conjoinct, dont pauvres humains las,
Voyans la sphere en la grand mer feconde
Estre comprise eurent joye et soulas
Et de la mort eviterent les lacz,
Considerans, soubz tel similitude,
Que Vierge et mere, en toute sainctitude,
Contiendroit Dieu et que, par son octroy,
Seroit de grace en concept si eprise
Qu'on la diroit hardiment, sans effroy,
Just astrolabe où la phere est comprise.

[35] See Bennett, *The Divided Circle*, 14–16; and A. Turner, *Early Scientific Instruments: Europe 1400–1800* (London, 1987), 11–16; for the planispheric astrolabe and its workings see H. S. Saunders, *All the Astrolabes* (Oxford, 1984).

used in surveying and to measure altitude; in navigation it could be used to determine geographical latitude and the direction of true north. Additionally, it played an important role in casting horoscopes, for it allowed astrologers to reference an "accurate" map of the heavens for a given time or locale.

The front of the astrolabe has a raised circumference called a limb. This is inscribed with a degree scale (usually a scale of hours); on the interior of the limb (the mater) a universal astrolabe can be fitted with alternative plates that depict the night sky as stereographically projected onto the plane of the equator from different latitudes. An observer selects a plate according to latitude and fits it into the mater. By this means a system of celestial coordinates based on the reference point of the observer's horizon is established. Overlaid above this celestial template is a second projection: a moveable skeletal or open-plan star map, known as the rete or spider, which points out prominent stars and the ecliptic of the sun. The rete can be rotated above the latitude plate to mimic the daily apparent motion of the celestial sphere. The changing positions of the stars in altitude and azimuth can be charted by reference to the plate lying below it. On the other side of the astrolabe is the alidade, a pivoted arm lying across the back face of the instrument. This has small sighting holes (the pinules) that are raised above the astrolabe's plane. If the instrument is held vertically by the shackle and ring at the top or throne, the alidade can be used in conjunction with a fixed scale that runs along the outside rim of the instrument to measure the height of a given object. For example, to establish a model of the night sky all that is required, at least in theory, is to measure the altitude of one of the stars on the rete by using the alidade and degree scale on the back of the astrolabe, then to adjust the star represented on the rete to the appropriate altitude lines. By following this method, the time, as well as the positions of other celestial bodies, can be determined.

Yet despite these putative practical uses, the cumbersome nature of handling the astrolabe in the field, often in less than perfect conditions, combined with limitations imposed by its necessarily small size (usually between 10 and 40 centimetres), the lack of textual corroboration of instances of use, as well as the precious nature of the materials utilized in its construction (typically gilt brass), make it extremely unlikely that the astrolabe was used for anything except as a showcase item for display or for didactic purposes in teaching the principles of astronomy and geometry. Nevertheless, according to Crignon's poem it was by using an astrolabe that the intrepid sailor, "lost on the ocean's expanses without anchor, sail or mast," could navigate to "a safe and good port."

Despite his assurances, however, the astrolabe he describes in his chant was not a navigational instrument. This is not to say that astrolabes were not used for navigation, but rather that such astrolabes were entirely different from the one described by Crignon in his poem. Appropriately, the navigational instrument was called a mariner's astrolabe.[36]

The first reference to the mariner's astrolabe coincides with the expansion of Atlantic shipping toward the end of the fifteenth century. More closely resembling a simple quadrant or a theodolite turned on its side than the astrolabe pictured in Crignon's verse, the mariner's astrolabe was a practical instrument particularly well-suited to long-distance oceanic navigation.[37] It has very little in common with its namesake.

Mariner's astrolabes were usually made of brass or iron, with the plates, inscriptions and decorative paraphernalia that could catch the wind and interfere with functionality removed. Essentially, the mariner's astrolabe was a heavily weighted circular ring. Its limb was inscribed with a degree scale; the alidade swiveled around and across this divided circle. To navigate a ship, sun or star would be sighted through the pinholes on either side of the alidade. A traveller would take readings of the sun on successive days at its highest point as it crossed his meridian. The angular distance could then be determined by comparison with measurements of the horizon; this, adjusted with appropriate declination tables, could then be used to find a ship's approximate latitudinal bearing.[38]

This contrasts markedly with the detailed craftsmanship and intricately inscribed plates of a planispheric astrolabe, which in theory could be used in surveying, navigation, time telling and target finding with artillery, but in reality was not.[39] Given all this, why did Crignon deliberately model his poetic astrolabe on a device that was never used for navigation? The

---

[36] See A. Stimson, *The Mariner's Astrolabe: A Survey of Known Surviving Sea Astrolabes* (Utrecht, 1988); D. Waters, *The Sea or Mariner's Astrolabe* (Coimbra, 1966); Bennett, *The Divided Circle*, 33–34; Turner, *Early Scientific*, 65–68; G. Beaujouan and E. Poulle, "Les origines de la navigation astronomique au XIVe et XVe siècles," M. Mollat and O. de Prat (eds.), *Le navire et l'économie maritime du XVe aux XVIIIe siècles* (Paris, 1957), 112–113. There are very few surviving mariner's astrolabes. In the Oxford Museum for the History of Science, for example, which has the largest collection of astrolabes in the world, there is only one mariner's astrolabe.

[37] See J. Law, "On the Methods of Long-distance Control: Vessels, Navigation and the Portuguese Route to India," in J. Law (ed.), *Power, Action and Belief: A New Sociology of Knowledge?* (London, 1986), 234–263.

[38] See Chapter 5 for a more detailed exposition of sixteenth-century navigational techniques.

[39] As Thomas Blundeville notes in his *Exercises, containing sixe Treatises … which Treatises are verie necessarie to be read and learned of all yoong Gentlemen that … are desirous to haue knowledge as well in Cosmographie, Astronomie, and Geographie, as also in the Arte of Nauigation…* (London, 1594): "Broade astrolabes, though they be thereby the truer, yet for that they are subject to the force of the winde, and thereby ever moving & unstable, are nothing meet to take the altitude of anything, and especially upon the sea; which this to avoid the spaniards doe commonly make their Astrolabes or Rings

apparent mistake was surely deliberate, for there can be no question that Crignon, an experienced navigator and explorer, knew the difference. Indeed, in his chronicle of Parmentier's voyage to Sumatra he took pains to note that he measured the height of the sun at midday almost every day.[40] There is little likelihood that he did this with a cross-staff, which was best suited for observation of celestial objects less than 45 degrees above the horizon and was certainly not a tool one would want to use for naked-eye solar observation at high noon.[41] That Crignon used a mariner's astrolabe can be plausibly deduced not only from his journal, but also from a poem written by his captain, Jean Parmentier. Read before the Puy at approximately the same time as Crignon delivered his *Just Astrolabe Where the Sphere Is Comprised*, Parmentier's chant, *The Mapemonde of Human Salvation*, described how difficult it was for a "cosmographe" to use an astrolabe to sight the pole star as his ship approached the equator:

> The wind behind him, he follows the North star,
> His chart in hand, he navigates
> Beneath the zenith of his own (the Northern) hemisphere,
> By measuring, in parallel lines
> The mapemonde of human salvation.
> But for this pilot, by whose reckoning
> The seas are crossed, his star reclines
> So that, for the elevation
> Of the pole to see, the astrolabe inclines.
> But the height of the bear [Ursa Minor] is concealed
> Just as it is sighted, its light disappears
> By retrograde (motion), in quadrature [in the shape of a square]
> Such that the sailor has no idea, by knowledge, craft or ruse
> How to pilot his ship, except to always have hope
> That its light will give him a glimmer of
> The mapemonde of human salvation.[42]

---

narrow and waightie, which for the most part are not much above five inches broade, and yet doe waigh at least foure pound, and to that end the lower part is made a great deale thicker than the upper part towards the ring or handle. Notwithstanding most of our English Pilots that bee skilful doe make their Sea Astrolabes or rings sixe or seven inches broade, and therewith very massive and heavie, not easie to be moved with everie wind." Regarding these difficulties, see Bennett, *The Divided Circle*, 34; and D. Howse, "Navigation and Astronomy," *Renaissance and Modern Studies* 30 (1986): 62–3.

[40] For details, see Chapter 5.

[41] Though Jean Rotz, Crignon's contemporary from Dieppe, claims to have done so. See Chapter 5.

[42] Parmentier, *Oeuvres*, 25.

> Le vent arriere, il suyt du North l'estoille,
> La chart au poing, se conduysant par elle
> Soubz le zenith de son propre hemispere,
> En compassant, en ligne parallelle,
> […]

It thus seems obvious that we attribute Crignon's "slip" to poetic license. The complexity of the planispheric astrolabe clearly provided a much richer technical vocabulary for Crignon's verse than did the bare-bones navigational instrument. Nevertheless, more is going on here. To modern eyes it might appear that the association of the mariner's astrolabe with the planispheric astrolabe was a means of grounding the abstruse and difficult theories that mathematicians and astrologers employed in such eminently practical uses as navigation.[43] Yet in the early modern period this gesture to a rhetoric of utility might have done more to undercut the efficacy of the navigator's craft than to bolster it.[44] Though counterintuitive, it was just the opposite move that needed to be made: navigators and mathematicians had to steer clear of the monsters of the contingent world in favor of more abstract universally recognized truths. An important step in this process was to associate the working navigational instrument with its more lavish and vastly more complex cousin, the planispheric astrolabe.

Planispheric astrolabes were relatively esoteric instruments in the early sixteenth century. Few had experience with them; fewer still possessed Crignon's recondite knowledge of their complex structure and use.[45] Navigation in the North Atlantic was at this time a kind of craft knowledge learned through long experience in well-known and frequently traveled waters, having more to do with body-to-body transference of skills

> A ce routtier que par dimensïon
> Mers traversoit, son estoille recline
> Tant qu'il convient, pour l'elevatïon
> Du polle voir, que l'astralabe incline.
> Mais les haulteurs du plaustre concellées
> Apperceust lors, ses clartéz reculées
> Par retrograd, en quadrature telle
> Que le routtier ne scayt art ou cautelle
> Pour pyloter, fors que tousjours espere
> Que de lueur luy donnera scintelle
> La mapemonde aux humains salutaire.

[43] See, for example, Neal, "The Rhetoric of Utility."

[44] Natalie Zemon Davis makes a similar point regarding commercial arithmetic: "honorable association for business was not conceived in terms of the fruits of commerce and finance. Rather business lost its stigma because business arithmetic was allegedly a liberal art and somehow related to the discovery of "great secrets and high mysteries." N. Zemon Davis, "Sixteenth-century Arithmetics on the Business Life," *Journal of the History of Ideas* 21 (1960): 18–48, 29.

[45] His poem pre-dates the vernacular texts having to do with the astrolabe's anatomy and use, for example, Dominique Jacquinot's *L'usaige de l'astrolabe* (Paris, 1545) and Jacques Focard's *Paraphrase de l'astrolabe contenant: Les principes de la geometrie, la sphere, l'astrolabe, ou déclaracion des parties de la terre* (Lyon, 1546). There were, however, a number of Latin texts produced in the first quarter of the sixteenth century that might have been available to the merchant poets of Dieppe, as for example, Ioannis Martini

honed by constant practice than with books, elaborate instruments or mathematical abstractions.[46] It required knowledge of the tides and of landmarks; familiarity with birds, fish, kelp and water conditions; and experience in the use of a sounding line, a compass and perhaps a portolan chart. The use of new navigational instruments and the mathematical techniques associated with them were relatively well known on the Iberian peninsula in the early years of the century.[47] But this knowledge diffused slowly. Thus, for example, across the channel, and as late as 1575, "even such relatively simple equipment as the cross-staff, the mariner's astrolabe, and the plane chart had been used aboard English ships for just a decade or two at the most."[48] Navigation in the early sixteenth century was a menial craft carried out by practical men in familiar waters. There was nothing particularly learned or valorizing about it. At the same time, the loss of a ship and its crew could have devastating financial consequences for those who had invested in it. This lent impetus to efforts to ensure the safety and dependability of transatlantic travel through the development of new mathematical techniques of navigation based on universal theoretical principles and not on dangerously contingent local knowledge.[49] Such techniques had to be "marketed" as necessary and useful tools to ensure the profitability of overseas trade. At the same time, as we shall see, this marketing also needed to appeal beyond mere economic interests, to social, intellectual and spiritual ambitions as well.

Poblacion's *De vsu astrolabi compendium* that saw no less than 5 Parisian editions between 1500 and 1527; Johannes Stöffler's *Elucidatio fabricae ususque astrolabii*, also published in 4 editions between 1512 and 1524 in Oppenheim, and a 1525 edition in Augsburg, and Laurent Fries, *Expositio vsusque Astrolabij* printed in Strasbourg in 1522. Using – and understanding – astrolabes was generally a hands-on affair. Paper instruments might have served the pedagogical needs of university students, but it was far different than using the instrument to navigate on the slippery decks of ships at sea.

[46] See E. H. Ash, *Power, Knowledge, and Expertise in Elizabethan England* (Baltimore, 2004), 89; see also Chapters 3 and 5.

[47] Waters, *The Sea or Mariner's*, 16. This is not to say that there was not resistance amongst Spanish pilots to the introduction of mathematical techniques of navigation; see A. Sandman, "Mirroring the world: Sea charts, navigation, and territorial claims in sixteenth-century Spain," in P. Smith and P. Findlen (eds.), *Merchants and Marvels: Commerce, Science, and Art in Early Modern Europe* (New York, 2002), 83–108.

[48] Ash, *Power, Knowledge, and Expertise*, 139.

[49] See ibid., for example, 96, 103, 139. Not coincidentally, the development of new navigational techniques was paralleled by the development of new commercial techniques of maritime insurance and underwriting. Indeed, not long after Crignon produced his chant praising the many virtues of the astrolabe, the first book published in France on maritime insurance, the *Guidon, stile et usance des marchands qui mettent a la mer*, was published in Rouen; see Mollat, *Le commerce*, 393; and L. A. Boiteux, *La Fortune de mer. Le besoin de sécurité et les débuts de l'assurance maritime* (Paris, 1968); see also, Chapter 5.

Navigating an Audience

In writing his poetic astrolabe for the Puy de palinod, Crignon was playing to an audience of actual and potential patrons. In addition to well-known courtiers, participants in the Puy included some of the most celebrated citizens of Dieppe and Rouen, including members of the old nobility, high-ranking members of the robe, wealthy merchants and important ecclesiastical officials. Not only was the cream of Norman society present when Crignon and Parmentier read their poetry, including such luminaries as Pierre Monfauld, the president of Normandy's parlement; Louis Cannossa, Bishop of Bayeux and correspondent of Erasmus; and Clément Marot (France's most illustrious, and infamous, poet), but also important merchants, ship-owners and civic leaders, such as Jehan Bonshoms, Gillebert le Fevre and Pierre Couldray. These men provided crucial financial and logistical support for overseas trade and exploration at a time when the French king was far more concerned with Milan and Naples and his rivalry with Charles V than he was with the transatlantic interests of Normandy's merchant community.[50]

These merchants traded in wool, silk, spice, alum, and brazilwood, used to make brilliant red dyes.[51] In the first half of 1529 alone, over 200 tons of the *bois de braise*, as brazilwood was known, was brought into Rouen's port.[52] However, the poetry read out before the assembled members of the Puy was far more than an attempt to garner the patronage of Normandy's merchant humanists for their skills as merchant navigators. It was also a form of symbolic address through which Crignon and Parmentier articulated the interests of, and associated themselves with, like-minded courtly and provincial elites who could help them on their journeys – not only across unknown seas – but also across social hierarchies, from lowly men of the sea to laurel-crowned poets.

Vernacular poetry in the first half of the sixteenth century played an important role in the French court with the likes of André de la Vigne, Jean Lemaire de Belges and Jean Marot writing verse and orchestrating

---

[50] See, for example, A. C. Vigarié, "France and the Great Maritime Discoveries–Opportunities for a New Ocean Geopolicy," *Geo-Journal* 26 (1992): 477–481.

[51] See Wintroub, *A Savage Mirror*, esp. chapter 2; on the importance of this trade see also M. Mollat, *Histoire de Rouen* (Toulouse, 1979), 154; Mollat, "Anciens Voyages Normands au Brésil," *Bulletin de la Société de l'histoire de Normandie* 5 (Rouen, 1887–90), 236–239 and 249–267; E. Gosselin, *Documents authentiques et inédits pour servir à l'histoire de la Marine Normande* (Rouen, 1876), 142–171; and M. Desmont, "Le Port de Rouen et son commerce avec l'Amérique," *Société Normand de géographie* 33 (1911): 404–410.

[52] See Mollat, "Anciens voyages," 257.

royal ceremonial to immortalize their royal patrons.[53] This gave poetry and its exponents both status and prestige at court and in the provinces.[54] The valorization of poetry, and the persuasive power of eloquence it was thought to embody, paralleled the growth of new urban centers of culture and commerce and the civically minded bourgeoisie that ruled them. The linguistic expertise that they cultivated went hand in hand with increasingly centralized bureaucratic forms of fiscal, military and juridical administration.[55] It is therefore not surprising that at the same time that poets were composing verse to honor princes and kings, they were also participating in religious poetic confraternities organized by the educated, urbane and cultured elites of provincial cities, such as Rouen, seat of Normandy's parlement and, after Paris, the second-largest city in France, and the important commercial port of Dieppe.[56]

The poetic confraternities of Rouen and Dieppe were at the center of Normandy's spiritual and cultural life. The Puy de palinod was among the most important of these. Each year on December 8, the day of the Immaculate Conception of the Virgin Mary, a competition would be held where *facteurs* (agents) of the Puy would read their poetic compositions dedicated to the Virgin. Dieppe's poetic concourse met not long afterward, on August 15, at the l'église Saint-Jacques to celebrate the Virgin's Assumption. Both confraternities were organized around mastery of complex poetic forms as detailed by *Le Grant et vray art de pleine*

---

[53] See, for example, F. Joukovsky, *La gloire dans la poésie française au XVIᵉ siècle* (Geneva, 1969), and Anne-Marie Lecoq, *François Iᵉʳ imaginaire: symbolique et politique à l'aube de la Renaissance française* (Paris, 1987).

[54] It is important to note that whereas Aristotle might have distinguished between rhetoric, grammar, and poetry, the tendency of sixteenth-century humanists was to conflate them. See, for example, E. Rummel, *The Humanist–Scholastic Debate in the Renaissance and Reformation* (Cambridge and London, 1995); and P. Kristeller, "The Modern System of the Arts," in idem, *Renaissance Thought II: Papers on Humanism and the Arts* (New York, 1965), 163–227.

[55] See, for example, Wintroub, *A Savage Mirror*, and Timothy Hampton, *Literature and Nation in the Sixteenth Century: Inventing Renaissance France* (Ithaca, NY, 2001); Walter Mignolo, *The Darker Side of the Renaissance: Literacy, Territoriality and Colonization* (Ann Arbor, MI, 1995); and Benedict Anderson, *Imagined Communities* (London and New York, 1983), esp. 37–46.

[56] There were also notable poetry confraternities in Amiens and Caen, for example. Regarding the Puy de palinod, see Wintroub, *A Savage Mirror*, esp. 64–90; D. Huë, *La poésie palinodique à Rouen: 1486–1550* (Paris, 2002); Dylan Reid, "Patrons of Poetry: Rouen's Confraternity of the Immaculate Conception of Our Lady," in A. van Dixhoorn and S. S. Sutch (eds.), *The Reach of the Republic of Letters: Literary and Learned Societies in Late Medieval and Early Modern Europe*, 2 vols (Leiden, 2008), I: 33–78, G. Gros, *Le poète, la vierge et le prince du puy: Etude sur la poésie mariale en milieu de cour aux XIVᵉ etˣⱽᵉ siècles* (Paris, 1992); C. B. Newcomer, "The Puy at Rouen," *Publications of the Modern Language Association of America* 31 (March 1916): 211–231; E. de Robillard de Beaurepaire, *Les puys de palinod de Rouen et de Caen* (Caen, 1907); G. Lebas, *Les palinods et les poètes dieppois* (Dieppe, 1904).

rethorique written specifically for this purpose by the Rouennais priest and poet, Pierre Fabri. The contributions submitted for the Puys' competitions demonstrate an unusually high degree of technical skill and expertise in the writing of verse. An invitation to the annual competition of Rouen's Puy printed in 1516 explains: a chant royal should contain *"xj lignes pour chacun baston sans coupes feminines silz ne sont synalimphées."*[57] In other words, each stanza (*baston*) of a chant royal was to have 11 lines, with caesurae (*coupes*) structured into each line placed after the fourth masculine syllable (unless the tonic final "e" was elided (*synalimphées*), thus making the syllable feminine).[58] The 1533 invitation went even further, stating that submissions had to be "well written, with correct orthography, diphthongs and grammar; otherwise, they … will be rejected."[59] In terms that Pierre Bourdieu has made familiar, the cultivation of arcane and highly technical grammar by the Puy's poets demarcated a field of expertise that distinguished them as part of France's new bureaucratic–civic elite.[60] As Pierre Fabri aptly put it, *"rhetoric donc est science politique."*[61] Indeed, for him, rhetoric was a science of "royal nobility" (*noblesse royale*) and "magnificent authority" (*de magnifique auctorité*).[62] He who possesses knowledge of it excels over all other men (*il a excellence sur les aultres hommes*).[63] The partipation of Crignon and Parmentier in Normandy's poetic confraternities was, in this sense, a performed act of social distancing – a means of distinguishing those who possessed the hard-won and recondite linguistic expertise to participate in the annual competitions of the Puy from those who did not. As Fabri put it, it was a means of discriminating between *"Sapiendum ut pauci en considerant la substance et signification"* of language, and *"loquendum ut plures en ensuivant le commun langage."*[64] Not only did these merchant-poets link their superior social status to spiritual justifications through the writing of Marian verse, they also

[57] Bibliothèque Municipale de Rouen, Ms. 1063 (Y. 16), fols. 1–2.
[58] Wintroub, *A Savage Mirror*, 69.
[59] BN, Ms. Fr. 1715, fol. 1ᵛ–2ʳ.
[60] P. Bourdieu, *Distinction: A Social Critique of the Judgement of Taste* (Cambridge, MA, 1984); Bourdieu, "Le langage autorisé: Note sur les conditions sociales de l'efficacité du discours ritual," *Actes de la recherche en sciences sociales* 5:6 (1975): 183–90; Bourdieu, "Social Space and the Genesis of Groups," *Theory and Society* 14 (1985): 723–744; Bourdieu, "Social Space and Symbolic Power," *Sociological Theory* 7(1989): 14–25. Also see G. Gadoffre, *La Révolution culturelle dans la France des humanists* (Geneva, 1997).
[61] P. Fabri, *Le grant et vray art*, I: 15. See also T. Reiss, *Knowledge, Discovery and Imagination in Early Modern Europe: The Rise of Aesthetic Rationalism* (Cambridge, 1997), 55–56.
[62] Fabri, *Le grant et vray art*, 9.
[63] Ibid., 7.
[64] Ibid., 13.

attempted to link it to the growing power of the state. Their poetic contributions to Normany's Puys paralleled the efforts of Geoffroy Tory, Lefèvre d'Étaples, and Guillaume Budé at court in attempting to transform the vernacular into a language suitable to guide the ship of state.[65] The reference to the king's instrument in Crignon's poem was thus meant in a double sense: first, to identify the king, François I (and France), with Norman exploits on the high seas, and then to link these exploits to the fulfillment of God's injunction to spread His redemptive Word throughout the world. Like the Virgin, the astrolabe was an instrument of God's benign will. By helping to make voyages across the seas possible the astrolabe also made possible the extension of God's Word across the earth, thereby fulfilling one of the preconditions for the prophesied end of the world and the thousand-year reign of Christ on earth.[66] As such the sphere of the astrolabe was not only a mirror of the heavens formed in God's own hand, but – as the guiding hand of the mother of God – a means of transporting all humanity from this world to the next. One would be transported, through its use, to "a port of grace," to "the year in which wars and all quarrels end" (*L'an où prend fin la guerre et toutz desbatz*).[67] Indeed, Crignon's poetic astrolabe figured, like the frise des sauvages at l'église Saint-Jacques, as a meditative device that could focus a supplicant's attention on the redemptive promise of Christian belief.[68] It was both a tool and a metaphor, an instrument that could be used to navigate the worldly seas, and a means of transforming earthly voyages into spiritual quests. Crignon concluded his poem by interlacing the voyages made possible by the astrolabe with those guided by the Virgin:

> Prince, the people had then certitude
> Of the port of grace and her longitude.
> And the pilot, suspending the astrolabe from his finger,
> Said: "Children, have no fear, I see
> The rays of the true sun break clear.
> This is the instrument that will ensure our safe conduct;
> The proof of which is in the using, and now I am sure,
> Just astrolabe where the sphere is comprised."[69]

---

[65] The measure of their success can perhaps be gauged by François I's 1539 edict of Villers-Cotterêts, which stipulated that French be the official language of the courts and in the administration of justice throughout the land.

[66] See Parmentier, *Oeuvres*, 46–48.

[67] Ibid., 64, see also 27–29, 40–41, 46–48.

[68] In this sense, Crignon's poetic astrolabe is perhaps similar to Charles de Bovelles's polyhedra; see P. M. Sanders, "Charles de Bovelles's Treatise on the Regular Polyhedra (Paris, 1511)," *Annals of Science* 41 (1984): 513–566.

[69] Parmentier, *Oeuvres*, 65.

Knights of Inscription

In the sixteenth century the planispheric astrolabe was an emblem of both mathematical and worldly power. Identified by John Blagrave in 1585 as the "mathematical jewel," it occupied a privileged status in the cabinets of kings, princes and nobles across Europe.[70] This owed as much to the intricate complexity and precious materials of its crafting as to its astrological uses, positing a close interweaving of earthly and celestial power. Seen from this angle, the conflation of the mariner's astrolabe with the planispheric astrolabe aimed to translate the ability to navigate the physical world into the ability to navigate the social one. Yet this act of social distancing was also an act of epistemic displacement whereby the image of the heavens inscribed on the astrolabe's brass plates was projected as a series of precise imaginary lines onto the globe, making possible the back-and-forth displacement of men, ships, and cargo across the open seas. Thus Crignon recounts how God inscribed the astrolabe with the tropics, the equator, the azimuth, and the almicanaratz with his "error-free compass," and how this would allow poor merchant sailors to navigate safely on the "vast and limitless seas." By conceptualizing the earth's surface as a geometrical grid defined by angular distance relative to the heavens, Crignon was also navigating an escape from local empirically based craft knowledge to the highly abstract "universal" framework of transoceanic navigation.[71] So important was this mathematical projection that Crignon began his account of his many voyages with a detailed exposition of how the earth was to be inscribed with lines of latitude and longitude:

In order better to understand lands and their relative locations and distances, we must know what constitutes the longitude and latitude of regions. According to cosmographers, longitude is reckoned from the meridian of the Canary Islands eastward along the equator until, the earth having been encompassed, the said meridian is again reached. According to modern navigation, as established by the Portuguese, this circle is divided into 360 degrees, each 17 leagues in length. This is true for the equinoctial as well as for the longitudinal line. Latitude refers to another imaginary circle, crossing at right angles the equinoctial line passing through the two poles, and encircling the earth ... These lines of longitude

---

[70] J. Blagrave, *The Mathematical Ievvel: shewing the making, and most excellent vse of a singuler instrument so called: in that it performeth with wonderfull dexteritie, whatsoeuer is to be done, either by quadrant, ship, circle, cylinder, ring, dyall, horoscope, astrolabe, sphere, globe, or any such like heretofore deuised: ... The vse of which iewel, is so aboundant and ample, that it leadeth any man practising thereon, the direct pathway ... through the whole artes of astronomy, cosmography, ... and briefely of whatsoeuer con- cerneth the globe or sphere ...* (London, 1585).

[71] See Ash, *Power, Knowledge, and Expertise*, 90.

or latitude extend over the surface of the earth – the latitude being determined from the elevation of the pole, or from the altitude of the sun; and the longitude by [the positions of] the moon and the fixed stars, or by the eclipses, or by even more subtle means unknown to many.[72]

This grid framework was far from imaginary. It had practical effects in making overseas trade possible. In Aristotelian terms, it helped translate unpredictable contingencies experienced on the open seas into memories durably inscribed on paper, vellum and brass, helping to transform voyages of exploration into normalized trade routes. This "imaginary" demarcation was celebrated with ritual solemnity. Thus, on the morning of May 11, 1529 Crignon recounts how, as they sailed toward Sumatra, some 50 men were made knights (*chevaliers*) for having passed beneath the equator (*passant sous l'equateur*). As was appropriate for this "*feste de chevalerie*" the assembled crew sang the Mass of the Blessed Virgin and dined on albacore and bonito.[73] It was not only the earth that was inscribed with knowledge of the heavens, but in chivalrous rites of passage, the men who sailed beneath them on the ocean seas.

Through these men, these knightly agents of translation, the inscription of the heavens on the earth extended across not only the globe, but also across the various and diverse inhabitants of the world. As Crignon said, the Tupinamba of Brazil are "like a blank canvas to which a brush has not yet been applied and on which nothing has yet been drawn, or like a young colt which has not yet been broken."[74] In this case, the *tabula rasa* of the New World's naked savages was to be covered with Christ's redemptive words as written and spoken by Normandy's merchants and sailors in the name of the French king and in the language of proto-nationalist dynastic competition between France and Portugal.[75] Crignon lent further legitimacy to practical mathematics, beyond appeals to utility, by plotting mathematics, exploration and trade along the axes of humanist linguistic virtuosity and Christian eschatology, connecting civic dimensions of humanism (the *vita activa*) to militant Catholic spirituality (e.g. Saint-Jacques as pilgrim and Matamoros) and to the social identities and professional activities of the merchant poets of the Puy. Not surprisingly, navigation and commerce, like the astrolabe, were poetically intertwined with an eschatological narrative of fall and

---

[72] B. G. Hoffman, "Account of a Voyage Conducted in 1529 to the New World, Africa, Madagascar, and Sumatra, Translated from the Italian, with Notes and Comments," *Ethnohistory* 10 (1963), 1–31, 33–79, 11–12 (from an account by "a great French sea captain from Dieppe," published by Ramusio in the third volume of his collection *Navigationi et Viaggi*, and generally attributed to Crignon). See Chapter 6.

[73] Crignon, *Oeuvres*, 19.

[74] Hoffman, "Account of a Voyage," 23.

[75] See Chapter 6.

redemption. Crignon's poetic confusion of the mariner's astrolabe with the planispheric astrolabe enabled him to intertwine his expertise in navigation and mathematics both with worldly (mercantile) power and with the very structure of the Christian cosmos. In another poem entitled *The Island Where the Earth Is Higher than the Heavens*, Crignon describes the paradise, found through divine cosmography and a "steady astrolabe and compass," where the dreams of Normandy's merchants for success and profit would be infused with profound religious significance:

> The great captain and wise pilot
> Who will sketch the great cosmography,
> Considering the many perils and passages
> That the sea holds, fortifies his great ship:
> His sacred vessel made quick and strong
> With the gifts of glory in immortal virtue,
> To lead her [his ship] to an island
> Where she could earn one-hundredfold,
> By which he made the crew joyful,
> For he promised with his map to reveal
> The island where the earth is higher than the heavens.
>
> On this island there is no wind or storm,
> The sea is temperate and serene,
> No taxes or imposts are to be paid,
> Azure, balms and spices abound,
> As do rubies, sapphires and fine gems,
> And precious gold of supernal value.
> It is the land where eternal power
> In human form desires to reign for all time,
> For in this place is the glorious Jesus,
> The heavenly empire, if we think about it right, is
> The island where the earth is higher than the heavens.[76]

---

[76] Ferrand, the modern editor of Parmentier's poetry, attributes this chant to Parmentier. But its placement in a series of chants royaux in BN, Ms. Fr. 379 by Crignon strongly suggests that this poem was authored by Crignon. It is so attributed in J. Nothnagle's collection of Crignon's works, *Oeuvres*, 81.

> Le grand patron et pilote tressage
> Qui composa la grand cosmographie,
> Considerant maint peril et passage
> Qui sont en mer, sa grand nef fortifie:
> Ses bortz sacrez, renforce et vivifie
> Des dons de gloire en vertu immortelle
> Pour la mener dedans une isle telle
> Qu'elle y pourra à cent doubles gaigner,
> Dont il a faict l'esquipage joyeux,
> Car il promet par sa carte enseigner
> L'isle où la terre est plus hault que ces cieulx.

Crignon's wager, his bid to transform his recondite knowledge of practical mathematics into creditworthy expert knowledge, and to garner patronage, prestige and authority, depended on a fourfold linkage between practical mathematics, humanist poetics, worldly power, and religion.

Translations

Whether charted across unknown seas or well-established social hierarchies, mobility in early modern France was reliant upon, and mediated by, expert knowledge and skill. With regard to the former, this knowledge might be instantiated in maps, instruments, ships, and stars; with regard to the latter, in gestures, clothing, tableware, speech, the building of a chapel, or the writing of poetry.[77] In either case, practiced familiarity with and expertise in navigational techniques and the norms definitive of social status were the *sine qua non* of mobility. At the same time geographical and social mobility were also made possible by undercurrents of spiritual belief. Whether as pilgrimage or as crusade, geographical distance, commerce, and colonization were subsumed into eschatological narratives of renewal, return, and salvation. Similarly, insofar as the early modern social world was dominated by ideas of the Chain of Being, movement up the social ladder was also movement closer to God. Geographic and social mobility were translated into theology and vice versa. The astrolabe of Crignon's poetry demonstrates the symmetrical translation of mathematical expertise into instruments capable of guiding ships across the seas and of navigating humanity toward redemption. From a slightly altered perspective it embodied the means by which the cosmographers, navigators, and pilots who guided these ships were translated into heroic spiritual athletes defending God and king in verse. From still further away, the poetic astrolabe can be seen as playing out a social gambit by

> Dedans ceste isle il ne faict vent ne orage,
> La mer y est temperee et serie,
> Payer ny fault coustume ne truage,
> Azur y croist, basme, et epicerie,
> Rubys, saphirs, et fine pierrerie,
> Or precieux de valeur supernelle.
> C'est le pays où puissance eternelle
> En corps humain veult en tout temps regner,
> Dont se en ce lieu est Jesus glorieux,
> Le ciel empire est à bien raissonner
> L'isle où la terre est plus hault que ces cieulx.
> ...

[77] See, of course, N. Elias, *Power and Civility* (New York, 1982).

translating the specialized craft knowledge of the navigator from a menial (*quadrivial*) form of expertise in practical mathematics first into poetry and then into politico-theological practice which intertwined practical mathematics, exploration, commerce, and colonization. This final linkage constituted something of a pathway around the jealously guarded knowledge of university theologians. The astrolabe symbolized the status and authority of the navigator and of the explorer. It embodied and conjoined theoretical and practical knowledge, helping make possible navigation on the open seas. By equating the astrolabe with the immaculately conceived Virgin, Crignon's chant can also be interpreted as a means of endowing the maritime exploits of Normandy's merchants with the legitimacy of a spiritual quest.[78] Crignon's poetry, like that of his captain's, thus aimed to link oceanic navigation not only with commercial profit, but also with the authority and prestige associated with veneration of the Virgin Mary. In attaching his mathematical knowledge to a ritual performance of verse, through the mediation of poetry and the status that it gave him, Crignon directly challenged those who would restrict access to Truth to those with formal university training in theology and natural philosophy.[79] The message of Crignon's poem closely resembles that of Oronce Fine, mathematics professor at the Collège Royale, when he argued in 1532 that mathematics participated in both natural and supernatural worlds and was the road to universal science – that is, to philosophy.[80] But Crignon's metrical translation of practical mathematics into verse aimed even higher up the early modern hierarchy

[78] On Normandy's merchants see Mollat, *Le commerce*, and, for a slightly later period, G. Brunell, *The New World Merchants of Rouen, 1559–1630* (Kirksville, MO, 1991).

[79] In 1542 the Sorbonne explicitly prohibited treatises of grammar, rhetoric, logic or *lettres humaines* from referring to Christian doctrine; see F. Higman, *Censorship and the Sorbonne: A Bibliographical Study of Books in French Censured by the Faculty of Theology of the University of Paris, 1520–1551* (Geneva, 1979), 50 and 52 n. 18. While it might be argued that the mixing of disciplinary genres such as poetry and mathematics violated the Aristotelian stricture against metabasis, from the fifteenth century there was a steady erosion of the borders demarcating these disciplines. Indeed, the skills, competencies, and local cultures associated with the status mobility of new professional classes were deeply implicated in the blurring of and in translations across the frontiers between different sorts of knowledge claims. See Westman, "Proof and Patronage," and Amos Funkenstein, *Theology and the Scientific Imagination from the Middle Ages to the Seventeenth Century* (Princeton, 1986), 36–37, 303–307.

[80] Orontii Finei Delphinatis, *Liberalium disciplinarum Professoris Regii, Protomathesis …* (Paris, 1532), fol. AA 2ʳ–AA2ᵛ, Aiʳ, quoted in Davis, "Sixteenth-century Arithmetics," 30. Such a claim could be made by Fine, who was a reader at the Collège Royale, precisely because he was dependent for his professional identity not on the University of Paris but on the patronage of the king. The hierarchy of the disciplines at the University clearly distinguished between such menial mathematical pursuits and true knowledge; trespass across disciplines was strictly forbidden. Normandy's Puys, however, offered opportunities to navigate across these divides.

of knowledge: beyond poetry, humanist learning, and philosophy to theology, the queen of all sciences.

Crignon's poem articulates a missionary evangelism whose agents were to be Normandy's merchant humanists acting in the name of God, good business, the public good, the king, and France. No longer were spiritual matters to be the sole purview of the doctors at the Sorbonne. Now poets, mathematicians, instrument-makers, merchants, and navigators could have their say as well.[81] Thus Crignon's chant was not just a poem, but a tool. It was a means of forging links and alliances between his technical expertise in practical mathematics, his cultural skills as a poet, the mercantile interests of his fellow poets, and the social legitimacy associated with a religious tradition of devotion to the Virgin Mary. His poetic astrolabe was not only a navigational tool, it was a spiritual device that was called upon to reveal and make evident the rational harmony imprinted upon the cosmos. It was also a rhetorical technology, a means to persuade his king – and his fellow poets – that overseas exploration was a legitimate, honorable and worthy undertaking. Furthermore, it was an instrument of social mediation whereby Crignon could attach himself and his knowledge to the status (and money) of powerful merchants, civic leaders and courtiers who could support his (and his captain's) endeavor to sail to the Indes orientales. Crignon's astrolabe was not only an instrument made of metal, it was one made of beautiful words.

[81] Crignon's poem was, in this sense, nothing less than an audacious intervention in the well-known controversy over the Virgin's Immaculate Conception by a lowly provincial navigator. As such, his poem was a direct challenge to traditional hierarchies of both knowledge and authority. On the controversies over the Immaculate Conception of the Virgin see E. D. O'Connor (ed.), *The Dogma of the Immaculate Conception* (Notre Dame, 1958); see also Wintroub, *A Savage Mirror*, Chapter 4.

3   Translation: Translating
       the Body of Thought

Translation is a common word. However, its meaning cannot be reduced to common usage: to turn one language into another. Made from the combination of two other words, *trans*, meaning to cross over or go beyond, and *fero* (whose supine form is *latum*), meaning to bear or carry, translation signifies movement and transference, transport, and carrying over. One of the earliest meanings of the word is to be found in rhetorical treatises where it was used as a translation of the Greek word for metaphor, that is, the use of words in such a way that they could travel beyond themselves – past their literal meaning – to signify something other than what they were typically taken to be. Quintilian, for example, begins his discussion of tropes in his *Institutio Oratoria* with this definition: *translatio* is a noun or verb that has been "transferred from the place to which it properly belongs to another where there is either no literal term or the transferred is better than the literal. We do this either because it is necessary to make our meaning clearer or ... to produce a decorative effect."[1]

In his discussion of metaphor in his important book, *European Literature and the Latin Middle Ages*, Ernst Robert Curtius turns to the practice of Roman poets, focusing his attention on the most commonplace of metaphors, the idea that the act of composition (or of reading) is like a ship set to sail: poets are sailors, and their minds, their ideas, their works, and their words are ships.[2] The metaphor of the ship is, of course, figurative, as all metaphors are, but it is also self-referentially literal; indeed, ships move, they transport, and they carry over – in other words, they translate. Robert Estienne alludes to this material notion of translation in his *Dictionarium Latinogallicum*, defining *translation* as taking someone's goods and giving them to another.[3] Similarly, one of the ways that the *Dictionnaire de L'Académie française* defined

---

[1] Marcus Fabius Quintilianus, *The Institutio Oratoria of Quintilian*, E. Capps, T. E. Page, and W. H. D. Rouse (eds.), trans. H. E. Butler. 4 vols (London, 1922), 3: 303.

[2] Ernst Robert Curtius, *European Literature and the Latin Middle Ages* (New York, 1953), 128–129.

[3] Estienne, *Dictionarium*, s.v. *translatio*.

translation was "the action of moving a thing from one place to another."[4] Ships, however, as Curtius's discussion of metaphor makes clear, did not just sail on the seas. The most famous carrying over of this nautical metaphor into the realm of politics can be found in Book VI of the *Republic* where Plato describes the fate of a ship after the mutiny of its crew. Each man considered himself capable of navigating the ship, and each vociferously proclaimed his right to steer. Yet despite these strongly held opinions, the crew was quickly distracted, turning its attention to the satisfaction of its gluttonous appetites in the ship's stores instead. On such a vessel, Plato laments, a true pilot "with knowledge of the skies, the stars and the winds, is [considered nothing but] a prater, a star-gazer, [and] a good-for-nothing."[5]

In the fifteenth century this ship of state earned a new name, the *Narrenschiff*, or "The Ship of Fools," by Sebastian Brant. Brant's *Das Narrenschiff* (1494) was a compendium in poetry and prose of the many different types of fool that comprised his ship's crew. Among the most relevant, in the present context, were those fools who "would innovate" and those who voyaged to distant lands to explore, describe and take the measure of the world and its peoples.[6] Only fools, he says, would wander to the ends of the earth in search of such knowledge while ignoring knowledge of their own mortal souls. It is perhaps ironic that Albrecht Dürer, an acknowledged master in measuring the world, was responsible for illustrating Brant's book.[7]

Like Plato's ship, Brant's *Narrenschiff* was less a satire than a didactic-moral admonition. In its mocking mirror, social and intellectual pretensions, immorality, and vice were exposed for what they were: madness and sinful folly that would lead inexorably to ruin, loss, and perdition.[8] There were a number of early French translations of *Das Narrenschiff*, for example, that by Pierre Rivière, based on Jacob Locher's Latin translation, published in Paris in 1497; a prose adaptation entitled *La petite Nef des Folz* by Jehan Droyn in 1498; and his more elaborate reworking in verse, *La grant Nef des Folz* published the following year.[9] Not long

[4] *Dictionnaire de L'Académie*, s.v. *translation*.

[5] *Plato's Republic*, trans. Benjamin Jowett (Cambridge, MA, 2008), 121.

[6] See Sebastian Brant, *La grant nef des folz du monde*, trans. Jean Drouyn (Lyon, 1499), fol. h ii[r]. See also, Christine Johnson, *The German Discovery of the World: Renaissance Encounters with the Strange and Marvelous* (Charlottesville and London, 2008), 38.

[7] Of course, Bosch also contributed to the popularity of the Narrenschiff motif. Charles D Cuttler, "Bosch and the *Narrenschiff*: A Problem in Relationships," *The Art Bulletin* 51:3 (1969): 272–276, at 272.

[8] See Peter Skrine, "The Destination of the Ship of Fools: Religious Allegory in Brant's 'Narrenschiff.'" *The Modern Language Review* 64:3 (1969): 576–596.

[9] See Dorothy O'Connor, "Notes on the Influence of Brant's 'Narrenschiff' outside Germany," *The Modern Language Review* 20:1 (1925): 64–70.

after these translations appeared, the Platonist physician and educator Symphorien Champier helped right the ships of fools, adapting the genre toward more explicitly positive models of virtue and heroism with publications such as *La Nef des princes & des batailles de noblesse* in 1502 and *La Nef des dames vertueuses*, the following year. Translation here shuttled between the political and the theological not as a negative counterexample, but as a meditative map – a guide – for crossing social and spiritual divides: thus were elites to be schooled and the righteous translated to paradise.[10]

As indicated by the titles of Champier's books, this movement was often expressed through the metaphor of *la nef* (the ship). The widely read book of sermons, the *Navicula Penitentie* (1511), by Brant's friend, the Strasbourg-based lay priest Johann Geiler von Kaysersberg, followed in this tradition, steering ships toward penitential, religious ends. The frontispiece, designed by Hans Burgkmair, depicts Geiler on the aft castle of his ship, preaching to its penitential passengers (Figure 3.1).[11] Jean Parmentier, in turn, transformed this penitential ship into an explicit vehicle of grace. Thus, in a poem he read before Dieppe's Puy de l'Assomption in the nef of l'église Saint-Jacques in 1527, he compared the Virgin Mary to a vessel of "sovereign beauty … [who had been] translated [*translatée*] to the sacred port where glory abounds."[12] More typically, the Virgin had already taken a permanent place in the heavens as the *Maris Stella* (Star of the Sea) – a beacon and a guide to crossing the world's perilous seas.[13]

The word for the central part of a church, the nave, derives, as we have seen, from the Latin for ship, *nāvis*. In French the two words were, and are, the same: la nef. From the present context, and as we saw in Chapter 1, it seems clear that the intertwining of these meanings in the nef – the secular, material ship on the sea, and the theological architecture of the body of the church – was not only related to the physical

---

[10] See, for example, Christine Hill, "Symphorien Champier's Views on Education in the *Nef des princes* and the *Nef des dames vertueuses*," *French Studies* 7:4 (1953): 323–334. In contrast to rutters and charts, maps were objects of worship, displayed on altars and in churches. See, for example, Flint, *The Imaginative*, 3–41. This recalls the meaning of translation as the "removal from earth to heaven, *orig.* without death, as the translation of Enoch; but in later use also said *fig.* of the death of the righteous": *OED*, s.v. "translation," n. I. 1c.

[11] Johannes Geiler von Kaysersberg, *Navicula penitentie* (Strasburg, 1512); interestingly, Mathias Schürer, the publisher, was also responsible for printing Konrad Peutinger's *Sermones conuiuales* (Strasburg, 1506), which included Fernandes' *De insulis et peregrinationibus Lusitanorum*, see Chapter 1 of the present work.

[12] Parmentier, *Oeuvres*, 28–29.

[13] Rubin, *Corpus Christi*, 178–198.

Figure 3.1 Hans Burgkmair, Woodcut frontispiece, Johannes Geiler von Kaysersberg, *Navicula penitentie* (Augsburg, 1511).

resemblance of naves (Figure 3.2) and ships, as some have conjectured, but to how both were intertwined in histories and practices of translation.

Naves and nefs both promised a journey; it would be mistaken, however, to think that the spiritual voyage promised by one was not part and parcel of the material voyage promised by the other. Indeed, spiritual, cultural, and commercial translations were entirely interconnected, whether

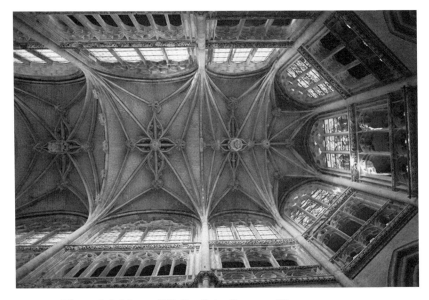

Figure 3.2 Nave of l'église Saint-Jacques, Dieppe.

on ships or in churches. Voyages of discovery and trade were conducted to carry the Word of God and win converts every bit as much as they were to find sources of gold, pepper, sugar, brazilwood, or slaves. Such voyages were, moreover, not only capable of carrying the Word (and along with it, goods) across the seas, they were also hermeneutic journeys across the book of nature, reading, or developing the capacity to read the Word as inscribed by God's own hand on the pages of the earth, the seas, the winds, and the stars. Geographical journeys were spiritual voyages, just as they were epistemic quests and commercial expeditions. For men such as Jean Parmentier and Pierre Crignon, social, economic, and spiritual aspects of translation were thoroughly conjoined. Naves and nefs were very much alike indeed. In the pages that follow, the manifold ideas and practices associated with translation will provide us with a kind of map by which to chart the voyages of Jean Parmentier across unknown seas, social hierarchies, the distant past, and his own unruly men.

Voyages and Conjurations

Parmentier set sail on Easter Day, March 28, 1529, on La Pensée along with her sister ship, Le Sacre, bound for the Indes orientales. Though an auspicious day, their luck was not good; Le Sacre, piloted by Jean's brother Raoul, ran aground and they had to delay their departure until April 2. Pierre Crignon's Log records that on that night a strange flame appeared in the heavens; he described it as being "round like a bowl out of which another flame shot illuminating the sky like sustained lightening."[14] With this sign noted, but not explained, La Pensée's voyage began.

They made good time, sighting Cape Finisterre on April 8. By the 17th they were making their way through the Canaries. On the 25th they reached the Cape Verde islands where they put in for food and water. On May 11 they crossed the equator, rounding the Cape of Good Hope a little more than a month later on June 22. Everything was going well and according to plan. While the crew went about its business guiding La Pensée toward Sumatra, their captain retreated to his library to work on his translation of Sallust's *War with Jugurtha*. The two projects were not unrelated.

The *Jugurtha* was by no means Parmentier's first work of translation – as understood not only as the movement of one language into another, but as a material carrying over from one place to another. Indeed, Parmentier was a sailor of renown, having captained – and sailed on – merchant ships that frequented Guinea, Le Terre Neuve, and Brazil, bringing back not only brazilwood (used to make a brilliant red dye), but a certain taste for the exotic.[15] The cargo of *La pèlerine*, a vessel that returned to Dieppe from Brazil not long after Parmentier's departure for the Indes, gives us a good idea of what was being brought over: "500,000 kilos of precious wood, 30,000 kilos of cotton … 600 parakeets already knowing some words of French, 3,000 skins of leopard and other animals, 300 monkeys, gold mineral, medicinal oils, all amounting to a value of 602,300 ducats."[16] Parmentier, moreover, translated this desire for goods from the New World into verse for Dieppe's Puy de L'Assomption. In his prize-winning chant royal, read at l'église Saint-Jacques in August 1527, he transformed the materiality of this worldly trade across the Atlantic into commerce with God, thereby lending a kind of spiritual legitimacy not only to the practical activities of merchant-sailors who voyaged to the New World, but also to the curiosity evoked by the exoticism of the goods they brought back. In his poem, French-speaking parakeets from Brazil became heavenly birds whose songs would greet the arrival (the "translation") of a merchant vessel into "the perfect port of welcome and joy":

[14] Crignon, *Oeuvres*, 13.
[15] See, for example, Guénin, *Ango et ses pilotes*, 126; and Crignon, *Oeuvres*, 95–110.
[16] See Paul Gaffarel, *Histoire du Brésil français au seizième siècle* (Paris, 1878), 77n1.

On the ocean and great worldly sea
Where without any assurance of repose
Was once a ship hastily launched,
Light and beautiful and good for all purposes:
To voyage to far off countries
And to carry a hold full
Of a rich red wood for the well-being
And grand profit of all humanity.
....
In this ship of sovereign beauty
There were not decrepit monkeys and marmots,
But parakeets from the heavenly realm
Who triumphantly spoke many beautiful words
To announce the moment she [the ship] was translated
To the sacred port where glory abounds.[17]

These commercial and poetic exploits were not the only works of translation Parmentier completed before he set sail for Sumatra in 1529. In 1528, for example, he translated Sallust's *De coniuratione Catilinae*, or *Bellum Catilinae*, for his patron Jean Ango (1480–1551).[18] As with his material (commercial) translations over the seas, Parmentier also wrote a theological translation in verse of Sallust's history. In his telling, Catiline and his co-conspirators were mutineers intent on destroying "*l'entreprise et les artz*" ("commerce and the arts") and defiling the "Temple of Divine Concorde."[19] In the concluding stanza of the poem, Parmentier explains his spiritual reading of Sallust's history: the "devil" and "vice," he says, were Catiline and his rebellion, while the Temple of Divine Concorde was

---

[17] Parmentier, *Oeuvres*, 27–29.

> Sus l'ocean et grosse mer mondaine
> Où jamais n'eust quelque asseuré repos,
> Fut jadis mise une barque soubdaine,
> Legere et belle et bonne à toute propos,
> Pour voyager en loingtaine contrée
> Et apporter une belle ventrée
> D'ung riche boys rouge par charité,
> Au grand profit de toute humanité
> ...
> En ceste nef de beaulté souveraine,
> Il n'y avoit vieilz cinges ne marmotz,
> Mais papegays de celeste demaine,
> Qui triumphoient de dire maintz beaulx motz,
> Et signamment quand el fut translatée
> Au sacré port où gloire est dilatée.

[18] Gaius Sallustius Crispus Sallust, *L'histoire catilinaire, composée par Saluste, hystorian romain, et translatée par forme d'interprétation d'ung très brief et élégant latin en nostre vulgaire françoys par Jehan Parmentier, marchant de la ville de Dieppe* (Paris, 1528).

[19] Parmentier, *Oeuvres*, 34–36.

the Virgin Mary. Despite his best (and most wicked) efforts, Catiline's rebellion failed. This, Parmentier argues, was because God's august power – as represented by the consul and the senate – foiled his attempts to sully the Virgin's purity and corrupt Divine Concorde. Sallust's history of the Catiline rebellion was thus used by Parmentier to imagine, in verse, a mysteriously Christian theocracy in ancient Rome.[20] Parmentier viewed his voyage, and his rule over his men, in similarly theocratic terms. Indeed, Sallust's histories, *The Catiline Conspiracy* and *The War Against Jugurtha*, played a central role in this journey and in La Pensée's mission of commercial, cultural, and spiritual conquest.

The Mind and the Body of Thought

In the *Catiline Conspiracy* Sallust recounts how Lucius Sergius Catilina raised an army and attempted to install himself and his partisans as rulers of Rome. The story Sallust tells, however, is not just about a failed rebellion, but about identifying the causes behind it.[21] Man, in his view, has a dual nature: on the one hand, he is a bestial being of the body, a thoughtless "slave to the belly" who uses his mouth only to eat; on the other, he is capable of suborning the body's base and avaricious desires and directing them toward virtue:

The purpose (the *puissance naturelle*) of the soul is to command and govern, and that of the body, to serve and obey…The soul bears a resemblance to God, which is in us the true seat of understanding, memory, reason and other divine virtues, while the body is like (*ayt commune*) a brute beast, which has only earthly sentiments. Therefore it would be more becoming to seek glory and honor with the virtuous help of the intellect (*esperit*) than by strength of arms, and other corporal things. But as the mortal life, which we are presently enjoying, is so fleeting, that if it were not prolonged by admirable memory dwelling with our descendants, we could hardly hope to aspire to glorious honor. For glory achieved by human riches, or by means of corporal virtues, is fleeting, frail, and fragile, with neither reason nor sense, but that which is won by virtue [of the soul or intellect] is illustrated by an infallible and eternal clarity.[22]

[20] Raymond Lebègue, "Un chant royal d'humaniste," *Bibliothèque d'Humanisme et Renaissance* 18:3 (1956): 432–435. It is interesting to note in this regard that just as Sallust's history was translated into theological verse by Parmentier, that the Roman historian, Sallust himself, had attempted much the same thing in transforming the ephemeral and contingent world of human action into case studies (i.e. the *Jugurtha* and the *Catiline Conspiracy*) that would reveal universal principles of progress and decline.

[21] See, for example, Erik Gunderson, "The History of Mind and the Philosophy of History in Sallust's *Bellum Catilinae*," *Ramus: Critical Studies in Greek and Roman Literature* 29:2 (2000): 85–126.

[22] Sallust, *L'histoire catilinaire*, fol. I^(r–v); see also Sallust, *The Histories*. Trans. J. C. Rolfe (Cambridge, MA, 1951), 3 (*War with Catiline* 1: 3–5). I would like to thank Corine Stofle for her help in translating this text.

However, Sallust continues, in the Rome of his day, the rule of the mind had been challenged, and indeed, overturned by ephemeral bodily desire:

[After they had] extended and augmented their lands, tamed kings, taken kingdoms, destroyed barbarians and cruel natives (*nativs*), and subjugated infinite peoples: and even laid waste to the mighty city of Carthage, their envious and great enemy, the Roman people ruled over all the world and peace replaced war. At that moment, fortune came as a blind fury to change everything for them, thus these Romans for whom excessive labor, terrible perils, and other austere things came fairly easy, fell into wicked idleness; they came to covet the riches of others which was for them a burden, a misery and a curse. From this execrable cupidity, greed and ambition for rule came the root and branch of all the ills afflicting the city of Rome. For where avarice reigns, faith, equity, and generally all good and virtuous qualities are subverted and destroyed: virtue was thus replaced by pride, insolence, cruelty, and contempt for divinity; everything had its price. And furthermore, the ambitious will to power compelled men to treason, treachery and dissimulation under the cover of words spoken with the sweet honey of gracious eloquence. But if one could see into their open hearts, and into the depths of their evil intentions, one would find but cruel deception and a bitter land of a most dangerous consequence, be they friend or be they foe; … because their friendship is not otherwise founded than on the particular benefit hoped to derive from satisfying their desires; and as to their enmity, one can surmise a similar judgment. *And don't think that you know them perfectly from the front they affect, for though they may smile, beneath the well-worn mask, is a lying heart.* These two damned and execrable vices Avarice and Ambition, began to grow, little by little, … until the city was overwhelmed and intoxicated, *as though by a contagious pestilence…* And the empire and government, which before had been so fair, so gracious, and so good became cruel, tyrannical, and intolerable.[23]

Recalling another, now mostly forgotten, meaning of the word translation, "the transference of a disease from one person or part of the body to another,"[24] Sallust's history evokes a colonial nightmare where the conquered take revenge on their conquerors by infecting them with a "deadly plague." He goes on to give further specificity to the source of the contagion: it is not imperialism per se – for example, the "taming of kings" or the destroying of "barbarians and cruel natives" – but the conquest of Asia that was responsible. Sallust explains: when Lucius Sulla led his army into the "voluptuous lands" of Asia, he allowed his men to forget themselves and become slaves to their bodies, luxuriating in drunkenness, whores, and all their "libidinous desires" such that the warlike spirit of Rome's soldiers was overcome by the "wicked and effeminate customs of sensuous love."[25] It was, Sallust concludes, contact with the decadent

---

[23] Ibid., 17–18 (*War with Catiline* 10:1–6), italics added.
[24] See the *OED*, s.v. translation.
[25] Sallust, *L'histoire catilinaire*, fol. IX^r; Sallust, *The Histories*, 21 (*War with Catiline* 11:5–6).

East that was responsible for corrupting the *virtù* of Rome; "riches, luxury, greed and pride united to invade, ravish and consume the young soldiers of Rome. Forgetting about their forefathers, they turned instead towards the desire for the goods of others and without either shame or modesty … abused all that was human and divine."[26]

Germs and Money

Asia was the source of contagion; it was also where Parmentier was headed in search of gold and spice. Though a very different geographical entity in the early modern period, Sallust's description of the decadence and luxury of "those charming and voluptuous lands," and their corrupting influence on the Romans, must have been troubling for Parmentier. Commerce in his day was frequently viewed as a danger to both social and moral order. He thus made it clear that he was not going to Asia for want of money, but rather for God, king, and country.[27] Nevertheless, the similarities between Sallust's descriptions of a world where men would rather show a good front than a good heart, and his own France, were striking.[28] As his contemporary, the anonymous Norman *farceur* and author of *Moral de tout le monde* (ca. 1535), lamented:

> The World by ambition
> Calls itself noble who hasn't any rents
> And by mad presumption
> Does not recognize either kith or kin
> The world puts honor up for sale…
> Today, all the World has gone mad.[29]

In the spring of 1529 when Parmentier set off for Sumatra, all the world must have indeed seemed mad.

---

[26] Sallust, *L'histoire catilinaire*, fol. IX^r; Sallust, *The Histories*, 21 (*War with Catiline* 12:2). See also Barbara Weiden Boyd, "*Virtus Effeminata* and Sallust's Sempronia," *Transactions of the American Philological Association* 117 (1987): 183–201.

[27] Parmentier, Oeuvres, 92.

[28] Sallust, *L'histoire catilinaire*, fol. VII^v–fol. VIII^r; see also Sallust, *The Histories*, 17–19 (*War with Catiline* 10:1–6).

[29] Émile Picot, *Recueil général des sotties*, 3 Vols (Paris, 1902–12), 3:43.
> Le Monde par ambition
> Se dict noble, qui n'a pas rente,
> Et par folle presomption
> Descongnoyt parent et parente.
> Le monde mect honneur en vente;
> …
> Qu'aujourd'hui Tout le Monde est fol.

See also Wintroub, *A Savage Mirror*, 134; Heather Arden, *Fools' Plays: A Study of Satire in the Sottie* (Cambridge, 1980), 148.

Few decades were as cruel as the one beginning in 1520. There were severe outbreaks of the plague in 1522 and again in 1527. The winters were unusually cold. Food was scarce and the land was unyielding. And while wages were low and kept falling, prices were high and kept rising.[30] The poor left the countryside and boiled into the cities of Rouen and Dieppe; begging and crime were rampant. Not surprisingly, the response of city officials was ferocious.[31] Even so, the poor rarely threatened social order. Rather, the problem was the nobles. Having squeezed their peasants for money rents and labor services, they still came up lacking, so they turned upon themselves. David Nicholls points to the seeming paradox: "the relative poverty of the ruling class seems … to have caused more trouble than the possible misery of the subordinate class."[32] Despite monarchical attempts to discipline the nobility, intranoble violence in the sixteenth century was endemic.[33] Land rich and cash poor, they were, moreover, vulnerable to incursions from below – by men who could trade money and knowledge for land, power, and title. This was a two-way street, however, as increasing numbers of nobles derogated their status and authority by stooping to trade and commerce. The reputation of the nobility suffered for all these reasons, and it was not uncommon for them to be portrayed as committing a kind of treason against the very foundations of social order.

The greed and avarice of the nobility, it was argued, were undermining the very basis of Christian community. A reflection of this attitude can be discerned in the pressure – exerted by virtually every quarter of French society – to formulate "laws governing derogation and the incessant *recherches de noblesse*."[34] Urban and religious rituals in Normandy translated these social ideals into festive performance and critique. In

---

[30] On the economy of upper Normandy in the sixteenth century, see, for example, Guy Bois, *The Crisis of Feudalism: Economy and Society in Eastern Normandy c. 1300–1550* (Cambridge, 1994); Michel Mollat, *Le commerce,* passim; David Nicholls, "Social Change and Early Protestantism in France: Normandy, 1520–62," *European Studies Review* 10:3 (1980): 279–308, esp. 281–283; Philippe Goujard, *La Normandie aux XVIᵉ et XVIIᵉ siècles: face à l'absolutisme* (Rennes, 2002), 36–70; and Philip Benedict, *Rouen during the Wars of Religion* (Cambridge, 1981), esp. 1–46.

[31] For example, see Gustave Panel (ed.), *Documents concernant les pauvres de Rouen, 1224–1634* (Paris, 1917), 20–21.

[32] Nicholls, "Social Change," 285.

[33] Carroll, *Blood and Violence.*

[34] Gayle Brunelle, "Narrowing Horizons: Commerce and Derogation in Normandy," in Mack Holt (ed.), *Society and Institutions in Early Modern France* (Athens, GA, 1991), 63–79, and idem., "Dangerous Liaisons: Mésalliance and Early Modern French Noble Women," *French Historical Studies* 19:1 (1995): 75–103; Gaston Zeller, "Une Notion de caractère historico-sociale: La Dérogeance," *Cahiers internationaux de sociologie,* n.s. 22 (1957): 40–74; and Etienne Dravasa, *"Vivre noblement": Recherche sur la dérogeance de noblesse du XIVᵉ au XVIᵉ siècles* (Bordeaux, 1965), 70–85.

the *Farce nouvelle moralisee* (c. 1461), for example, an actor portraying Le Monde (the World) proclaimed in no uncertain terms that nobles, far from being servants of the "*royaulme de France*," were "a heap of lascivious rogues, scoundrels, and thieving pilferers."[35] But as the rest of the title of this farce (*Des Gens Nouveaulx qui mengent le monde et le logent de mal en pire*)[36] makes clear, the nobility was not the only culprit singled out for bringing about this state of affairs. The play's anonymous author places the blame for the World's madness, and for its going "from bad to worse," squarely on the shoulders of the *gens nouveaulx* (the new men) rather than just the nobles, that is, the lawyers, functionaries, physicians, merchants, artists, writers, and poets who knew years before Francis Bacon did that knowledge was power.

Though many suffered during the hard pestilential years of the 1520s,[37] the new men, and in particular those involved in Normandy's maritime trade, flourished.[38] Indeed, while the nobility was selling out and marrying down, and peasants and urban poor were starving, the new men were doing quite well. These *gens nouveaulx* could not claim noble blood, but they had money and much-needed knowledge. Moreover, unlike the nobility, they had nothing to fear in stooping to commerce and trade or medicine and law. It is thus perhaps not surprising that they were much resented as corrupters of social and moral order, having a reputation for being little more than scheming parvenus who used their ill-gained expertise in rhetoric, law, money, and commerce to rise to stations for which they had no right and no qualifications other than their own self-proclaimed virtue.[39] As the author of *Le Moral de tout le monde* lamented:

> All the world wants to be noble,
> Even the most villainous, such as they are.
> Each says himself to be a gentleman,
> Were he more foul (more common) than a dead rat.[40]

[35] Picot, *Recueil général*, I:28: "Ung tas de paillars, meschans, coquins, larrons, pillars."
[36] Ibid.: *Of the New Men Who Devour the World and Take It from Bad to Worse*.
[37] There were subsistence and mortality crises in 1522 and 1531: see Emmanuel Le Roy Ladurie, *The French Peasantry, 1450–1660*. Trans. A. Sheridan (Berkeley, 1987), 102 and 10–12, 106; Bois, *The Crisis*, esp. 384–385; Henry Heller gives a nice summary in *Labour, Science and Technology in France, 1500–1620* (Cambridge, 1996), 28–30.
[38] See, in particular, Mollat, *Le commerce*, 523–541.
[39] See, for example, Wintroub, *A Savage Mirror*, 124–141; Arden, *Fools' Plays*, esp. 138–159; and Jean Alter, *Les origines de la satire anti-bourgeoise en France: Moyen âge-XVIᵉ siècle* (Geneva, 1966), 156–217.
[40] Picot, *Recueil general*, III:36.

> Tout le Monde vault ester noble,
> Tant de race villain soyt il.
> Chascun se dict ester gentil,
> Fust il plus villain c'un rat mort…

New Men: Then and Now

It is fascinating to think that Parmentier had just left this world on his way to Asia when he began translating Sallust's *Jugurtha*. Though he sailed on La Pensée, it surely did not escape his attention that the name of his brother's ship, Le Sacre, connoted not only the holy rite of consecrating a church, the coronation of a prince or prelate, or a bird of prey that annually travels – according to Nicot – south "toward the Indes,"[41] but also, at least according to Cotgrave, "a greedie fellow, one that makes boot of all he can lay his clutches on; also an excessive glutton, or gully-gut, and a spendall, unthrift, squanderer, extream rioter (especially in respect of his bellie)."[42]

The text Parmentier was translating on board La Pensée makes the parallels between Rome and sixteenth-century France clear. In the *Jugurtha* Sallust turns from the site of infection in Asia to explore its disastrous effect on the body politic. In the first instance he seems to argue that the decline of Rome was entirely the fault of the unrestrained greed and ambition of the nobility. Writing as a new man, Sallust argues that the nobility constructed its honor and status from words rather than deeds, from imposture and deceit, rather than virtuous action. This, in any case, was the opinion professed by the speech Sallust wrote for Gaius Marius:[43]

Compare me now, fellow citizens, a 'new man,' with those haughty nobles. What they know from hearsay and reading, I have either seen with my own eyes or done with my own hands. What they have learned from books I have learned by service in the field; think now for yourselves whether words or deeds are worth more.[44]

A little farther along, Marius appropriates and transforms the embodied soul of noble identity, the *imagines* – the effigies and masks – of one's ancestors, into a bodily manifestation of his social and political legitimacy. He thus translates the meaning of the effigy from reverence for the deeds of the dead (and, of course, from the notion of inherited nobility) into renown for virtuous deeds done in the present:

I am not able to display, for the sake of your confidence, imagines nor triumphs or consulships of my ancestors, but, if matters require it, I can display

---

[41] Nicot, *Le Thresor*, s.v. *sacre*.

[42] Cotgrave, *A Dictionairie*, s.v. *sacre*.

[43] *Novus homo* is defined by Badian in the *Oxford Classical Dictionary* as follows: "A term used in the late Republic (and probably earlier) for the first man of a family to reach the Senate … and in a special sense for the first to obtain the consulate and hence nobilitas." See, for example, D. R. Shackleton Bailey, "Nobiles and Novi Reconsidered," *The American Journal of Philology* 107:2 (1986): 255–260; and Timothy Wiseman, *New Men in the Roman Senate, 139 B.C.–14 A.D.* (Oxford, 1971).

[44] Sallust, *The Histories*, 313–314 (*War with Jugurtha* 85:13–16).

spears, a banner, decorations, other military honours, and even further, scars on the front of my body. These are my imagines, these my nobility, not left by inheritance, …but things which I myself sought, through my many labours and dangers.[45]

Sallust, like Marius, was a self-identified new man, but he was less confident and far more circumspect and ambivalent than the words he gave to Marius. The new men, he realized, were by definition ambitious social climbers. Part of the blame for Rome's descent into decadence and discord, thus belonged to homines novi. The trope of the greedy and avaricious new man was hardly an invention of late medieval and early modern France; *farceurs*, such as the author of *Le Moral de tout le monde*, echoed commonplace classical critiques of the homines novi. As Sallust put it: "even the 'new men' who in former times always relied upon worth to outdo the nobles, now make their way to power and distinction by intrigue and open fraud rather than by noble practices."[46] By contrast, Sallust proposes that though true nobility might, of course, be based on military exploits or political service, it might also consist in acts of writing. Sallust, in this sense, displaces family history with history as a vocation. In contrast to Marius, who looked to his own self-made history as a kind of charismatic *imago* to legitimize his position, Sallust viewed his written histories of Rome – and his practice as a historian – as masks themselves: as imagines made of words rather than captured spears, banners, or scarred flesh.[47]

For Sallust, history was not just a description of Rome's decline; neither was it an epidemiological exercise aiming to elucidate causes. Rather, it was therapeutic; while diagnostic, it offered prescriptive cures. Public life, he argued, had become so corrupt and venal that retreating from it, and writing its history, was the sole means left to him to slow Rome's seemingly inevitable downfall.[48] Like the imagines of one's ancestors, "that kindled in the breasts of noble men [a] … flame that [could not] … be quelled until they by their own prowess … equaled the fame and glory of their forefathers,"[49] he expected that his histories would stand as exempla of virtue and reason to counteract the inimical effects of luxury, avarice, vice, and ambition.[50]

---

[45] Ibid., 317 (*War with Jugurtha* 85:29).
[46] Ibid., 139 (*War with Jugurtha* 4:7–8).
[47] See Lauren Kaplow, "Redefining *Imagines*: Ancestor Masks and Political Legitimacy in the Rhetoric of New Men," *Mouseion* 3:8 (2008): 409–416.
[48] For example, Sallust, *The Histories* 7 (*War with Catiline* 3:1) and 137 (*War with Jugurtha* 4:1).
[49] Ibid., 139 (*War with Jugurtha* 4:6).
[50] Ibid., 137 (*War with Jugurtha* 4:3).

Some sixteen centuries later, Jean Parmentier had a similar view of words and their importance: for him, it was the act of translation itself that was to have salubrious effects. As he told Ango in the dedicatory preface to *De coniuration*, this text will be of great help in the work of wisely running the city of Dieppe and will be of great public utility.[51] "Man," he tells Ango, "is made for the reason and support of man, such that it is the urgent affair of every individual to look after every other."[52] "But," he continues, "this sort of amiable and virtuous society has always been – and still is – difficult to find."[53] It is my "fantasy," he says, that this translation might help you to repulse the age-old villains of "avaricious particularity" and thus maintain Dieppe as an amiable and virtuous community.[54] For this reason, he suggests an act of translative cannibalism: "Such things" (Sallust's history), he says, "are worthy of being well digested."[55] In this view, in this fear, and with this advice, Parmentier shares much with the man he was translating. As with Sallust, so too for Parmentier: the intellect and culture – the mind – were to be cultivated as a propaedeutic to just rule.[56] These ideas would be sorely tested as he rounded the Cape of Good Hope.

A Sacred Thought

On the first day of July the weather began to change. The good seas they had enjoyed since leaving Dieppe were gone. The torments were so great that Crignon noted in his journal that he believed that the gods of the wind, Aeolus and Favonius, were dancing, along with Affricus Libo and Thetis, to celebrate their nuptials: "They were accompanied by many great fish … assembled in great troops leaping and parading about, while we in our ship, we were also dancing up a storm on the high seas."[57] A few days later, as they were passing through the Islands of Fear, La Pensée entered upon its opposite: the Sea without Reason. At this point, Crignon reports, the body of the ship was "weary, disappointed, and

---

[51] Parmentier, *Oeuvres*, 120: "Fruict de grande utilité et d'exemple bon et salutaire à la chose publique" ("Fruit of great utility and an example of good and salutary governance.")

[52] Ibid.

[53] Ibid., 118.

[54] Ibid., 119.

[55] Ibid., 120: "sont dignes d'estre bien digerées."

[56] Petrarch embraced this comparison between history and the *imagines* of the ancient Romans. See Petrarch. *Epistole familiari*. Ed. V. Rossi. 4 vols. (Florence, 1933–42), 2:80, cited by Patricia Osmond, "'Princeps Historiae Romanae': Sallust in Renaissance Political Thought," *Memoirs of the American Academy in Rome* 40 (1995): 101–143 at 106.

[57] Crignon, *Oeuvres*, 23.

frustrated" as well as being plagued by "scurvy, … fever, cramps, purple blemishes that covered their legs and thighs, as well as those other maladies that so many of them had gained by their merit before leaving home, like syphilis and sores of the groin, about which," he timidly adds, "I will say nothing."[58] Frightened and resentful whispers turned to talk of mutiny and a speedy return to Dieppe.

Parmentier responded not with force per se, but with words. Crignon recounts the scene:

> observing that several of his followers were angry and weary at being at sea for so long, and lamenting the lives of ease they had left behind, he [the Captain] composed a little Oration on the marvels of God and the nobility of man, to give them heart to carry on and complete their voyage, in which they had previously served so well.[59]

Through his words Parmentier hoped to draw his men toward higher things, and thus distract them from their fears and their mutinous desires.

He began with a question – the same question that was surely plaguing his men – "What am I doing here?":

> I ask myself, for what fantasy
> I quit Europe; and why I dream so much
> That I want to behold all Africa.
> And still more, why I think that I would not be satisfied
> If I could not pass beyond the ends of Asia,
> ………………
> Where did this reckless desire come from?
>
> Should I say, with Horace or Juvenal,
> ………………
> That I go to the Indes to flee poverty?
> This argument is false and malformed:
> Lack of money cannot hurt me,
> ………………
>
> To what purpose am I therefore resolved
> When I conceived of a voyage so arduous?[60]

Reason comes to Parmentier's rescue and responds:

> When the desire moved you
> To take on this curious trouble
> This you did, for sake of honor

---

58 Ibid., 29.
59 Parmentier, *Oeuvres*, 4.
60 Ibid., 91–92.

As the first Frenchman to undertake
A voyage to a land so faraway
....
Your enterprise is for the glory of the king
For the honor of your country and for yourself.[61]

Here seems to be the answer: for king, country, and honor. But this does not satisfy Parmentier. Overcome by melancholy, he retreats to his library to find comfort in his books, in particular, the holy scriptures. Reason alone, it seems, could not provide a satisfactory answer to his question; he was in need of divine guidance to assuage his doubts and continue his voyage. Thus, he turned to "the Ecclesiastique" (the Book of *Ecclesiastes*) for solace. Listening to his words, he realized that he had undertaken this journey not for king, country, or gain, but for God, for "he who wants to have great glory and honor, must follow God, his sovereign Seigneur."[62]

But how was one to follow God? Theology, of course. But theology presented difficulties of its own. How could a man like him, a sailor, and expert in transporting men and goods on the sea, attain to the heights of theological knowledge, the queen of all the sciences. As he put it:

I have no money to put
On the table to have benefices;
And without money, one has neither commissions nor letters of mark
Unless one is shrewd enough to undertake

---

60 *(cont.)*   Je suis pensant pour quelle fantasie
Je quicte Europe; et tant je fantasie
Que veulx lustrer toute Affriqe la nove.
Encores plus, je ne me rassasie
Si je ne passe oultre les fins de Asie,
...
Mais dont me vient telle effrenée estude?
Diray je, avec Horace ou Juvenal,
...
Que aux Indes voys pour fuyr povreté?
Cest argument est faulx et anormal:
Faulte d'argent ne me peult faire mal,
...
Sur quel propos suis je donc arresté
Quant j'ay conceu voyage si pesant?

[61] Ibid., 92.

'Quand ce vouloir te esprit,
De te donner tant curïeuse peine,
Cela tu feis, afin que honneur te prit
Comme françoys, qui, premier, entreprit
De parvenir à terre si loingtaine;
...
Tu l'entreprins à la gloire du roy,
Pour faire honneur au pays et à toy.'

[62] Ibid.

> A theft: by guile or by flattery.
> But I have had other propitious means
> To find honor as others have,
> In acquiring the solemn round bonnet of a doctor at the University.[63]

For Parmentier, knowledge of sun, stars, the winds, and the sea, and knowledge of ships and of men – all his know-how, both tacit and practiced, and all his nautical expertise – was as worthy of authority and admiration as the "solemn round bonnet of a doctor at the University." There is more than an echo of Marius here as he continues:

> Because, one takes for a good pilot,
> A sailor, whose skill all have recognized,
> Who is understood to be well practiced,
> And who has known his profession for a long while,
> So perfectly that he never fails by so much as an iota,
> That a doctor is not always in the university.
> Am I not therefore so filled with certainty
> To have elected seafaring to study
> To leave the gentle for the rude?[64]

At first, Parmentier seems entirely unwilling to accept traditional intellectual and social hierarchies that devalue hands-on practical experience in favor of training in philosophy and theology. But then he begins to question his vocation as a sailor, wondering aloud whether or not it was blocking his way to true knowledge and authority. Thus he imagines abandoning the rude for the gentle to become a doctor:

---

[63] Ibid., 94.

> Car aussi bien, je n'ay argent à mettre
> Sur le bureau, pour avoir benefices;
> Et sans argent, on n'a bulle ne letter
> S'on n'est subtil pour s'entremettre
> D'en crocheter, par dol ou par blandices.
> Mais je eusse eu aultres moyens propices
> D'avoir honneur, comme les aultres ont,
> In acquerant le grave bonnet rond.

[64] Ibid.

> Or, pour certain on tient qu'un bon pillotte,
> Un marinier, qui tout son cas bien note,
> Bien entendu et bien exercité,
> Est plus long temps pour entendre sa note,
> Parfaictement qu'il ne s'en faille ïote,
> Que un docteur n'est en l'université.
> Suis je pas donc bien plein de cecité
> D'avoir eleu le maritime estude,
> Laissant le doulx pour emporter le rude?

Consider what doctor I would have been
....
Instead, I am poor and dejected,
A Sailor without authority
Except when in danger while on the sea!
But would it be that on land, one could say to me: 'Our Master,
Bona dïes, your beautiful words, by Saint Gilles,
Are as true as the beautiful Gospel!'[65]

In the end, as Reason points out to him, his desire to speak words as true and beautiful as a saint was both prideful and arrogant; this was a role for which he had neither training nor skill. It would be better, says Reason, for him to pursue his vocation as a sailor, relying on his nautical expertise to follow a different course of study and matriculate into another, more expansive university, one where universals were to be the very stuff of the world – that is, the facts, the ephemera, the fleeting and contingent particulars so condemned by Scholastic philosophers and theologians. These, Reason proclaims, were God's works too, and knowledge of them was knowledge of – and homage to – Him.[66] Parmentier therefore advises his men to take advantage of this voyage to understand God's works – the dignity of man and the wonders of the seas, the earth, the air, and the heavens – and find in them happiness, peace of mind, and profit. He goes on to implore them to be virtuous and to take courage in God's grand design, explaining that it is in knowing how to read and interpret God's heavenly signs, as written in the complex motions of the heavens, that they would be able to sail to lands where they would find both wealth and glory:

---

[65] Ibid., 94–95.

> Considerés quel docteur je eusse esté,
> ...
> Et en lieu, suis un povre dejeté,
> Un mathelot qui n'a auctorité
> Fors qu'en la mer, quant au danger fault estre!
> Mais en la terre, on m'eust dict: 'Nostre maistre,
> Bonna dïes, vos beaulx mots, par Sainct Gille,
> Sont aussi vrays que la belle evangile!'

[66] Ibid., for example, 95. In this regard, Parmentier's *Oration* shares much with the views of his contemporary Paracelsus, who wrote that "it is the physician who reveals to us the diverse miraculous works of God… He should make it manifest; and not only what is in the sea, but also what is in the earth, in the air, in the firmament … so that many people may be able to see the works of God and recognize how they can be used to cure disease": Paracelsus, 67. Perhaps Paracelsus had read Parmentier's thaumaturgic oration in the posthumous edition edited and published by Pierre Crignon in Paris (1531), for his advice to the physician closely parallels Parmentier's proposed cure for the body of La Pensée and the advice he gave to his men.

By this voyage and navigation,
You know by experience [by sight]
That what I say is certain and true:
The body of the heavens proves it
What you see by elevation [i.e. by raising up an astrolabe]
Day to day in your latitude;
When the cold north where you make your home
You forsake and you apply yourself
To navigating toward the Mydi (South),
The way there is shown to you by the altitude [of the sun]
And thus your [home in the] North quickly sinks away,
Into the depths of the sea.
....
[Thus] as experts (*expertz*) you [will] see,
When you have [sailed across] half the span of the globe

... The marvelous evidence,
... the holy deeds and unmistakable signs
And the virtues of the one immortal King
Your grand lord, full of benign goodness,
True teacher of the divine sciences,
And your guide in this mundane world
....
[who if you follow well will deliver you, and]
You will find profit, joy and honor.[67]

[67] Ibid., 111–112.

Par ce voyage et navigatïon
Vous congnoissez en speculatïon
Ce que je dy par vraye certitude:
Les corps du ciel en font probatïon
Que vous voyez par elevatïon
De jour en jour en vostre latitude;
Quand le froit Nord où gist vostre habitude
Vous delaissez et mettez vostre estude
À naviguer vers la part de Mydi,
La part du Su[d] vous monte en altitude
Et vostre Nord descend en promptitude,
Tant qu'en la mer il est approfundy.
...
Lors, ce verrez, ainsi que gentz expertz,
Quand vous aurez du grand rond la moitié.
Ainsi voit on les merveilles insignes,
Ainsi voit on les haultz faictz, les grandz signes,
Et les vertus du seul immortel Roy,
Ton grand Seigneur, plain de bontés benignes,
Vray ensigneuer en sciences divines,
Et ton ducteur en ce mondain terroy.
...

There was clearly an aesthetic dimension to Parmentier's praise of God's works – their immensity, their beauty, their awesome and heart-stopping wonder – but his plea also had a more practical dimension. As he tells his crew: "By this voyage and navigation, you know … what I say is certain and true." It is through plying one's trade as a mariner whose experience of, and mastery over, nature in all its marvelous variety and sublime magnitude that one can find both true knowledge of God and "profit, joy and honor."

Yet his skills as a ship captain were not exactly impressing his crew. He therefore turned to poetry to ground and warrant his authority. Parmentier's *Oration*, in this sense, located La Pensée's journey within a narrative of redemption mediated by the poetic translation of the merchant-sailor's practical activities into theological verse. Poetry for Parmentier, like history for Sallust, was palliative – through it he could direct his crew to have faith not only in his skill as a captain, but in their expertise as sailors. These expert skills, he argues, were divinely sanctioned: knowing the world was knowing God.[68] Spurred on by their captain's oratorical and poetic power, his men could appeal to their much-maligned hermeneutic skills in reading the book of God's creation to succeed with their audacious enterprise for the "great profit of all humanity":

> Thus I concluded that it was the right moment,
> To escape ennui and despair,
> To write this oration in the small moments available to me
> As reason has brought me to hear
> And reading them to my crew so that they would understand.
> That all that I have done, of which God is the glory
> Who guides us and will guide us [back] again,
> As I pray, in making a successful voyage.[69]

> [who if you follow well will deliver you and]
> Tu obtiendras profit, joye et honneur.

Parmentier's poetry is often obscure and syntactically acrobatic. The phrase in brackets represents several lines that have been interpreted and summarized to create a comprehensible bridge to the final quoted line.

[68] See Smith, *The Body of the Artisan*, esp. 95–128.
[69] Parmentier, *Oeuvres*, 113.

> Par quoy conclus qu'il estoit bien saison,
> Pour eviter annuy et desplaisir,
> De escrire tout à mon petit loisir
> Comme Raison me avoit bien faict entendre
> Et à mes gens le donner à entendre.
> Ce que j'ay faict: dont à Dieu soit la gloire
> Qui nous conduict et conduira encore,
> Comme j'espere, en faisant bon voyage...

Parmentier read his poem to his men on the Pensée's rocking windswept decks as they sailed across the Sea without Reason. Like Sallust's histories, his *Oration* was an attempt to heal the body politic and restore civic and spiritual order to the ship of state. Perhaps his voice carried over to be heard on Raoul's vessel, Le Sacre. Together, sailing toward the Indes in search of spice, gold, and converts, the two ships formed a kind of material affirmation of the ideas Parmentier was trying to express in his oration; indeed, La Pensée and Le Sacre composed a sanctified thought capable of transforming the crews' suffering, fear, and anguish, as well as their greed and ambition, into a holy cause – into a quest.

Crossing Frontiers

It is difficult to imagine this scene, and even more difficult to imagine La Pensée's desperate men being at all persuaded by Parmentier's poetry, with its allusions to Horace and Juvenal, its 55 stanzas of rhyming verse, or its obscure and overworked biblical references. It is perhaps easier to imagine the crews' grumbling discontent, their hunger and thirst, their bodies wracked with pain and disease. Their humanist captain, with his classical erudition and his esoteric ideas about God, the heavens, and the world, surely did not inspire his hardened crew with confidence. Unlike Normandy's Puys, where he was addressing a like-minded audience of similarly learned and socially ambitious (new) men, here he was addressing an entirely different animal: one that was angry, frightened, poor, and uneducated. These were practical men who navigated using methods learned through tradition and experience, not the cosmographical works of Ptolemy, Sacrobosco, and Pierre D'Ailly. Though Parmentier wrote as if he and his men were brothers and compatriots, his *Oration* clearly points to their very different social positions and skills. Among men of the sea, some swabbed the decks and hoisted the sails, and others navigated using arcane and recondite theories and sophisticated instrumentation; in other words, while some worked with their hands and bodies, others worked (supposedly) only with their minds.[70] At the same time, however, like Sallust, Parmentier implored his men to be wary of those who like "pigs and lying swine" care only for the "pleasures of the belly,"[71] thus pointedly recalling the opening lines of the *Catiline Conspiracy*: "It behooves all men who wish to excel the other animals to strive with might and main not to pass

---

[70] On the dubiousness of claims regarding the theoretical underpinnings of practical navigation, see Chapter 2.
[71] Parmentier, *Oeuvres*, 100–101.

through life unheralded, like the beasts, which Nature has fashioned groveling and slaves to the belly."[72] In a curious and contradictory way, Parmentier seems to embrace both Sallust's valorization of the power of words to persuade, and Marius's valorization of the body as a kind of imago, insofar as it was through Parmentier's own bodily presence as an orator on the decks of La Pensée that he hoped and expected to persuade, much like a priest in the nave of a church delivering the Mass in Latin, a language all but incomprehensible to his congregation. The nave was where a congregation would assemble to listen to and be guided by the words of the priest. As we have seen, it was also the same word as ship in Parmentier's day. It is interesting to note how Hans Burgkmair played on precisely this double meaning in his frontispiece of Johann Geiler von Kaysersberg's *Navicula Penitentie*.[73] One can easily imagine Parmentier in the place of Geiler on board his ship. Parmentier, of course, was not a priest, and though he was speaking a language foreign to most of his crew, this language was still intuitively comprehensible to them in that it was a translation of their own working language of the sea.[74]

The reception of the *Oration* is surely a matter for speculation without corroborating evidence. What can be said, with some assurance, is that Parmentier was the captain and his men were not happy. Indeed, murder could not have been far from their minds. In this incendiary situation, Parmentier turned to his rhetorical skills as a means of steering his men toward the recognition of his authority. He did this by translating the skills of the seaman "*qui congnoissez la nature et les flotz*" ("who know nature and the tides) into knowledge of God.[75] Parmentier's

---

[72] Sallust, *The Histories* 3 (*War with Catiline* 1:1–3); see also Sallust, *L'histoire catilinaire*, fol. Iʳ.

[73] *Schiff*, which derives from Middle High German rather than Latin, retains the same conceptual reach as *nef* in referencing both ship and nave.

[74] In this sense, it is important to recall that Normandy (particularly the Pays de Caux), in the 1520s was a kind of seed crystal for the development of ideas associated with the reform. If Dieppe followed the demographic trends regarding the rise of Protestantism in the rest of France, then men such as Parmentier and his crew would have been among those particularly influenced by ideas having to do with direct worship, vernacular translations of the scriptures, and the challenging of priestly authority. On the rise of Protestantism in Normandy, see, for example, Gaston Le Hardy, *De l'histoire du protestantisme en Normandie depuis son origine jusqu' à la publication de l'Édicte de Nantes* (Caen, 1869); Henri Prentout, "La Reformation en Normandie et le debuts de la Reform à l'université de Caen," *Revue Historique* 114 (1913): 285–305; Charles Oursel, *Notes pour servir à l'histoire de la Réforme en Normandie au temps de François Iᵉʳ* (Caen, 1913); Nathanaël Weiss, "Note sommaire sur les débuts de la Réforme en Normandie (1523–1547)," *Congrès du millénaire normand* 1 (1911): 193–205; Victor Madelaine, *Le Protestanisme dans le pays de Caux* (Paris, 1906); Benedict, *Rouen During the Wars*, and Nicholls, "Social Change," and Goujard, *La Normandie*.

[75] Parmentier, *Oeuvres*, 97.

own power, in this regard, was reflected in the practices of his men. He was an expert in transporting ships and men back and forth across long distances precisely because he was an expert in the empirical world of particulars and ephemera, that is, the world of the body – of bodies – such as the men who composed his crew. However, at the same time as he was identifying his power with that of his men and the exigencies of the sublunary world, he made every effort to distance himself from the social origins of this power. Accordingly, he translated his, and his men's, tacit, experiential, and practical knowledge into theoretical, theological, and poetic terms of the universal; terms that would be all but incomprehensible to the lowly sailors who comprised his crew. By establishing this distance between the mind and the body, between his epistemic and moral authority and the desires of his crew, Parmentier set the stage for the rhetorical performance of his power to command the body of La Pensée.

Like the faithful translator that he was, Parmentier thoroughly understood the working native language of sailors and ships, but he also maintained his distance from them. As we have seen, Du Bellay, in his *La Deffence, et illustration de la langue Françoyse* (1549), likened such acts of understanding and distancing to cannibalism, in which the language to be translated was to be consumed, digested, and assimilated, and in the process, transformed into something new.[76] In this way, he argued, the French language could be renewed, invigorated, enriched, and perfected. This is precisely what Parmentier tried to do with the language of his men, transforming it into the specialized vocabulary of astronomers, cosmographers, poets, and theologians, thus giving it, and its object – the sublunary world of particulars – a new kind of credibility and authority.[77] A similar rhetorical performance was repeated a few weeks later, though in a far more dramatic, exemplary, and bloody fashion; it too was a display of Parmentier's authority, but this time as a literal – indeed, visceral – enactment of the distance between himself and his men, and between the mind and the body.

Cartographies of the Body and Maps for the Mind

On August 12, 1529, Pierre Crignon made the following entry into his journal:

On this day, the son of Pontillon, after having been sick for 2 or 3 months with some swelling on his head, died. And to know from whence this could have come,

---

[76] Joachim Du Bellay, *La deffence, et illustration de la langue Françoyse*. Ed. Jean-Charles Monferran (Geneva, 2001), 91.
[77] See, for example, Smith, *The Body of the Artisan*.

the captain performed an anatomy and cut his head open from ear to ear, finding there, on his brain, a huge cyst filled with corrupt and fetid matter and black blood, which stunk, and which had rotted through the bone of the head into its interior. Afterwards, he was laid to rest in the manner of sailors, God have his soul.[78]

The very next day a Breton sailor named Jean Dresaulx also dropped dead; he too

was opened by our captain to see why he was ill and what brought him to his end, and through this anatomy it was found that one of his lungs had decayed, and that his body was riddled with holes filled with red water turning yellow, and that he had a huge swelling filled with corrupt matter in the joint of his knee beneath the small bone.[79]

The brutality of these acts of public dissection was only thinly veiled by their putatively diagnostic aim: "to see why he was ill and what brought him to his end." One can only imagine how the crew must have felt about one of their own being cut open, eviscerated, and studied by their captain. Katharine Park has postulated that in Northern Europe the recently deceased had a liminal status in which soul and body were still in some sense conjoined. Corpses were treated as "active, sensitive, or semianimate."[80] In this context, she argues, to "open or dismember the body for doctors to inspect – an act of no conceivable utility to the deceased, now beyond all medical aid – was an act of objectification and a violation of personal honor."[81] It was also, in the present context, surely, a violation of the honor of the crew's collective body – that is, its corporate identity. For Parmentier, of course, it was not a question of the crew's honor at all, but about the honor of God. In this sense, the dissected bodies of Dresaulx and Pontillon, Jr. were constituted, as Park avers, by an act of objectification. They were, quite literally, like the sea, the stars, the skies, and the earth, another of God's works to be explored, understood, and mapped. Neither Parmentier nor Crignon distinguished between the physical site of the dissected body and other aspects of God's creation. In this sense, these dissections were like exclamation points on the message

---

[78] Crignon, *Oeuvres*, 31: "Cedit jour le fils de Pontillon, apres avoir esté malade deux ou trois mois de quelques apostumes qui luy estoient venues en la teste, mourut. Et pour connoistre don't cela luy pouvoit ester venu, le capitaine fit faire une anatomie et luy couper la teste tout à l'entour jusques aux oreilles; et luy fut trouvé sur la cervelle une grosse apostume pleine d'ordure et de sang noir, fort puante, qui avoit desjà pourri l'os de la teste par dedans. Apres il fut enseveli à la mode mariniere. Dieu en ait l'ame!"

[79] Ibid.

[80] Katharine Park, "The Life of the Corpse: Division and Dissection in Late Medieval Europe," *Journal of the History of Medicine and Allied Sciences* 50:1 (1995): 111–132 at 115.

[81] Ibid., 126.

of the *Oration*. Indeed, in the ship's journal, Pierre Crignon continues his description of the "opening" of the Breton sailor on August 12 to report matter-of-factly, without either pause or transition, that "on this day, the height of the sun at noon was 10 and a half degrees."[82]

In his book *The Body Emblazoned*, Jonathan Sawday sketches out the interconnectedness of such activities as anatomy and voyages of discovery in the first half of the sixteenth century. The anatomist provides a map of the body, just as the explorer maps out the lands he encounters on his intrepid journeys. As Sawday puts it, "the task of the scientist was to voyage within the body in order to force it to reveal its secrets."[83] Beyond the obvious anachronism, Sawday is right to point to the similarity between anatomy and transoceanic navigation, for both were implicated in practices now associated with histories of colonization and conquest. But it was not only the seas that needed to be navigated or unpredictable savages that needed to be tamed if La Pensée was to succeed on its mission. Its body too had to be subdued and directed by its mind – by reason – if mutiny was to be averted and the voyage toward the Indes was to continue. Nothing could better reinforce the locus of power on board the ship than the authoritative gaze of its captain opening up, examining, and analyzing his men's bodies to read the "signs" hidden there.

The exemplary violence of dissection, typically performed on criminals or outsiders without any links to local culture, family, or community, must have shocked and horrified La Pensée's crew.[84] Postmortem dissections were, as Foucault says, inscriptions of power on the body.[85] Writing on the dead took diverse forms; in this case, it was paralleled by an inquisition – a questioning of the body – to uncover, make manifest, and account for the circumstances of death. The corpses to be dissected, in this case, were not the accused, but the subject-victims who held hidden within clues left by the agents of their demise. Postmortem examinations were thus employed forensically to determine both cause of death

---

[82] Crignon, *Oeuvres*, 31.

[83] Jonathan Sawday, *The Body Emblazoned: Dissection and the Human Body in Renaissance Culture* (New York, 1995), 25.

[84] Katharine Park, "The Criminal and the Saintly Body: Autopsy and Dissection in Renaissance Italy," *Renaissance Quarterly* 47:1 (1994): 1–33. See also Michel Foucault, *Discipline and Punish: The Birth of the Prison* (New York, 1997), 3–69. These shipboard dissections may also have contributed to the mystique and authority of Parmentier's obscure and mysterious knowledge. There were similar diabolical associations with mathematics and mathematical navigation (see Chapter 2). In this sense, the so-called gaze of the anatomist had more than a little witchcraft and sorcery about it. Though these would seem to run counter to its legitimacy, the opposite was often the case, especially among nonelites. See, for example, Stuart Clark, *Thinking with Demons: The Idea of Witchcraft in Early Modern Europe* (Oxford, 1999), 458.

[85] See Sawday, *The Body Emblazoned*, 113.

and attribution of guilt. The body of a victim was the signifier and the signified, the evidentiary locus of a criminal act and the power of sovereign justice to write a historical-juridical narrative of proof – of guilt or innocence – and a justification for its monopoly over the legitimate use of violence. Parmentier similarly used his men's corpses as mediums of inscription to write upon and read off a narrative of his own social and epistemic power to direct and control the body of La Pensée.

Bodies and Minds

According to Parmentier's contemporary, the Neoplatonic professor of medicine and amateur mathematician Jean Fernel, the body consisted of three distinct regions,

> each fenced off, so to speak, by their own boundaries: the top one in the brain is the seat of sensation and reason; the middle one in the chest is the abode of spirit and life; the lowest one below the diaphragm and abdomen is the workplace of nature and the cherishing parts. In this way these parts that are soiled and fouled by filth have been banished by the providence of immortal God to the lowest position, the sump as it were of the body; and their removal to a distance prevents unpleasant vapors from contaminating and disturbing the heart and the brain, the seats of the main faculties and the senses themselves. This neighborhood is, so to speak, the palace kitchen, where some parts prepare foods like cooks, and get them ready for other parts that they are appointed to serve.[86]

What could be an anachronistic characterization of Parmentier's gaze as objectifying here takes on a more sixteenth-century resonance, for his epistemic stance toward his men closely corresponds to the social distance that was written into the very nature of human anatomy. Just as cooks in the kitchen live at a remove from and serve their superiors, so too did the crew of La Pensée – "the sump as it were of the body" – serve the ship's heart and brain, which is to say, its captain, Jean Parmentier. In his *Physiologia*, Fernel provides a detailed description of the scene of dissection, with the university professor and physician (the mind) directing the labor of his assistants and helpers (the hands) doing the anatomy, all the while providing a learned narrative, a natural history, for the spectators who were watching from a suitable distance away.[87] The role of the physician described by Fernel is precisely how Parmentier viewed his role as a ship captain and translator. It is also what Sallust had in mind for the writer of history.

---

[86] Jean Fernel, *The Physiologia of Jean Fernel (1567)*. Trans. John M. Forrester (Philadelphia, 2003), 65.
[87] Ibid., see 161. See also Steven Shapin, *The Social History of Truth: Civility and Science in Seventeenth-Century England* (Chicago, 1994).

According to Pierre Crignon's 1531 edition of Parmentier's works, Parmentier was a "good cosmographer and geographer" who "composed numerous globes and maps of the world, and many marine charts by which many have reliably navigated."[88] The goal of a good map, whether of the body, the skies, or the seas, was a sure and faithful translation. Parmentier describes just such a map in one of the poems he presented before the Puy in Rouen (1526). It was, he said, by using *La mapemonde aux humains salutaire* (*The World Map to Human Salvation*), the Virgin, untouched by original sin, that the cosmographer could plot a course to "Callicou to find mines of gold (and of course, paradise)":[89]

> With the wind behind him, he [the cosmographe] follows the
>   North star,
> His chart in hand, he maps out his course,
> Beneath the zenith of his own hemisphere,
> He navigates, in a straight line,
> The world map to Human Salvation.[90]

Similarly, one assumes, it was by reference to God's plan of the human body, and aberrations (particularities) located postmortem, that one could write a history of the causes of morbidity and plot a course toward an appropriate therapy. Like Sallust's *Catiline Conspiracy* and the *Jugurtha*, Dresaulx and Pontillon (the sailors dissected by Parmentier) were case histories that could reveal both the complex historical etiology of disease and also, putatively, the direction one should take to find a cure. As Parmentier stresses in his *Oration*, knowing the heavens was the means by which one could transport and lead both men and ships across vast distances. Similarly, it was by knowing (by opening) the bodies of his crew, that he would be able to translate the symptoms and empirical qualities of his men's diseased bodies into a story about causality, therapeutics, and his own unquestionable authority as the brain behind La Pensée. Finally, it was by knowing the internal working of language that he would be empowered – as the Norman priest, poet-grammarian, and rule-maker for the Norman Puys, Pierre Fabri said – to enflame "the lazy against all honorable perils" and to pacify "popular seditions and reduce all to good peace and tranquility."[91]

---

[88] Parmentier, *Oeuvres*, 4: "Mapesmondes en globe et en plat et maintes cartes marines sur lesquelles plusieurs ont navigué securement."
[89] Ibid., 24.
[90] Ibid., 25.

> Le vent arriere, il suyt du North l'estoille,
> La charte au poing, se conduysant par elle
> Soubz le zenith de son proper hemispere,
> En compassant, en ligne parallelle,
> La mapemonde aux humains salutaire.

[91] Fabri, *Le grant et vray art*, 6.

Figure 3.3 Manuscript illumination accompanying Jean Parmentier's poem, "Au parfaict port de salut et de joie" (c. 1528-1540).

This is precisely what Parmentier tried to do on board La Pensée when his men were threatening mutiny and perhaps even murder.

Parmentier's journey was made on the seas, but it was also made on social and epistemic topographies that were navigated through demonstrative displays of expertise, power, and authority. It was through this knowledge and expertise, underlined by the magisterial and pious authority he displayed in his *Oration* and the brutal force he deployed with his dissecting knife, that would ultimately define the success or failure of his voyage. It was by these means that Parmentier plotted his own – and his ship's – translation to what he called "the perfect port of welcome and joy" (see Figure 3.3).[92]

---

[92] Parmentier, *Oeuvres*, 27–29.

4   Scale: The Heart of the Matter

Fish have scales, so does justice; we can have scales upon our eyes, and we see them – scales – as they trap the ineffable sounds of music as notes on a page. Scale comes from the old French, *escalle* (from the Latin *scala*) for ladder. Ladders are, quite literally, scales. Estienne described scale as "*Le degree d'une eschelle.*"[1] One can climb a ladder's steps to the heavens; rise or descend the ladder of social hierarchy; dominate (scale) a mountain; be aroused to the heights of ecstasy (*climacter*: to climb, to scale); breach – *escalade* – the fortified walls of a castle, or move and inspire so as to inflame – as Estienne put it – "*publiquement quelcun, comme le mettre à l'eschelle.*"[2] This is surely what Parmentier was trying to do to his men on the decks of La Pensée when it was lost on the seas *sans raison*. Scale was also a way to measure – or rather to inscribe an instrument so as to make measurement possible; thus Chaucer described the circle on the "foreside" of an astrolabe as having a "marked" scale "in Maner of 2 Squyres or ells *in Manere of laddres.*"[3] This scale allowed its user to climb to the heavens and bring them down to the measure of man. The height of a building, a mountain or a star could be translated into a humanly graspable scale demarcated by an instrumental landscape of inscribed lines each of which corresponded to a number. Thus a perception of distance could be captured in brass and translated to even more mobile and reproducible media such as paper and vellum.

Scale could work in the opposite direction as well. Indeed, not only do maps make impossibly large distances present, they make these distances navigable. Put somewhat differently, scale was a means of disciplining the eye, of orienting the viewer of maps on a map, making possible the transposition of the self into the abstract space of a representation that literally allowed one to scale the world. Scale takes other forms besides movement up and down metaphorical ladders, however; it could also

---

[1] Estienne, *Dictionarium*, s.v. "escalle."
[2] Ibid.
[3] Geoffrey Chaucer, *A Treatise on The Astrolabe: addressed to his son Lowys*, Ed. W. Skeat (London, 1872), 7. My emphasis.

refer to weight and/or economic or spiritual value: that is, a scale was also an apparatus for weighing a commodity, the reliability of an adversary, the guilt of a criminal, or the souls of the dead.

Discipline and order are central components of these scales, whether commercial, juridical, or theological. Thus, for example, a less well-known definition of scale dating from the fourteenth century describes a scale as a squadron or battalion of men: a host of well-ordered and disciplined soldiers who as individuals come together to act in unison as parts of a larger organized whole. Overall size is irrelevant so long as there is an internal consistency as to the actions of a host's parts. Analogically, the connection to the careful ordering of inscribed lines marshaled on an astrolabe seem clear; more abstractly, one could point to the disciplining of the earthly particularities of geography and their organization into precise and unchanging relations relative to transformations of overall size on maps and globes; or a similar kind of relation with regards to architectural plans and models that disciplined materials and structures built to exacting requirements (that is, to scale), from the tiniest of models to the largest of buildings. Proportions had to be held constant, regardless of size.

Like expertise and translation, scale was also closely implicated in commerce – not simply in terms of its importance in composing weights and measures, but as steps, or sites where commerce happened; as Nicot said, an "*Eschelle, signifie encore, une place de commerce sur les costes de la mer Mediterranée.*"[4] Especially interesting here is scale's entanglement with the sea – that is, not only with scaly fish, but also with ships and merchants: as the *Dictionnaire de l'Académie Françoise* said, ships "Faire escale dans un Port, C'est y mouiller, y relâcher."[5] Or as Rabelais put it in *Gargantua*: "*Je retourne faire scale au port dont suis issu.*"[6] A scale was a step in a journey – a port of call, a temporary hiatus in the movement from one point and one port to another. And indeed, commerce itself has always been dependent on scales of value that could be transposed across borders, frontiers, and cultures. That these notions of mensuration were ontologically located on the physical sites of bodies should be of no surprise: fish and reptiles were perhaps the most immediate models, their minutely organized scales – overlapping, regular, and repeating – fit together as parts and as a whole by a structural coherence maintained across transformations of size, from small to large, and vice versa. One need not look to the sea, forests or deserts, however,

[4] Nicot, *Le Thresor*, s.v. *escale*.
[5] *Dictionnaire de l'Académie Françoise* (Paris, 1762), s.v. *escale*.
[6] François Rabelais, *Oeuvres de François Rabelais contenant la vie de Gargantua et celle de Pantagruel* (Paris, 1854), 58.

to find scales; scale was (and is) an ontological category, with man being the measure (the standard) of all things: an inch (the width across the knuckle of the thumb), a palm, a hand, a span (the distance from the end of the little finger to the end of the thumb of a spread hand); a cubit (the length of the arm from elbow to fingertip); a yard (the distance from the nose to the fingertips); a pace (the length of a stride).[7]

Bodily norms of measurement were applied not only to trade, but also to space, from the lived space of architectural form to the cosmos. For Vitruvius, the human figure was a mirror of divine proportion in which the rules governing classical architecture could be found. Following in his footsteps, men like Alberti saw in human proportion an accessible site of experimentation – a territory to be mapped – as a means of giving (practical) mathematical specificity to universal order. Thus, Alberti described a system of mensuration based on scaled relationships of the human form that could be derived empirically using an instrument resembling the mariner's astrolabe, a *finitorium* (see Figure 4.1).[8] Here the coherence of the human body was empirically mapped and used as the most immediate and central referent of precise relations describing the scale of the microcosm to the macrocosm.

A more metaphorical notion of mensuration can be seen in Parmentier's contemporary from Dieppe, Jean Rotz's *Book of Idrography*, where a Vitruvian man, not unlike Leonardo da Vinci's, is positioned as both the measure – and the measurer of – the heavens and the earth (see Figure 4.2). For Leonardo, the correspondence between the symmetry and proportion of man and the universe resonated with the geometric principles that he set out with pen and ink; Rotz's Vitruvian man, on the other hand, has been set free and endowed with the artist's powers to take the measure of the

---

[7] Taken from http://whitebard.tripod.com/append.htm. Recall Shakespeare's perhaps all too true parody of credit relations as embodied by the Jew, Shylock, and Portia's injunction that he accurately measure and weigh Antonio's flesh on pain of his own death:

> Therefore prepare thee to cut off the flesh.
> Shed thou no blood, nor cut thou less nor more
> But just a pound of flesh: if thou cut'st more
> Or less than a just pound, be it but so much
> As makes it light or heavy in the substance,
> Or the division of the twentieth part
> Of one poor scruple, nay, if the scale do turn
> But in the estimation of a hair,
> Thou diest and all thy goods are confiscate (William Shakespeare, *Merchant of Venice*, Act IV Scene I).

[8] Jane Andrews Aiken, "Leon Battista Alberti's System of Human Proportions," *Journal of the Warburg and Courtauld Institutes* 43 (1980): 68–96, at 75. See also Dalibor Vesely, "The Architectonics of Embodiment" and Robert Tavernor, "Contemplating Perfection through Piero's Eyes," in George Dodds and Robert Tavernor (eds.), *Body and Building: Essays on the Changing Relation of Body and Architecture* (Cambridge, MA., 2002), 28–43 and 78–93.

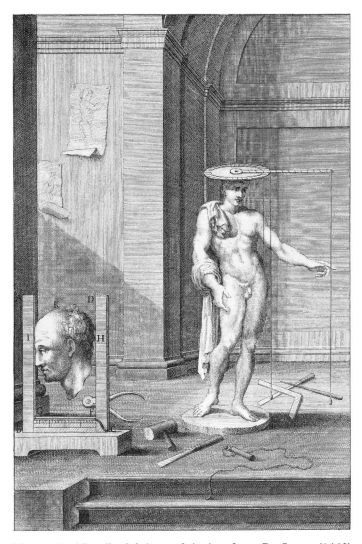

Figure 4.1 Alberti's *definitor* o *finitorium* from *De Statua* (1462), as depicted by Pio Panfili's engraving for *Della architettura della pittura e della statua* (Bologna, 1782).

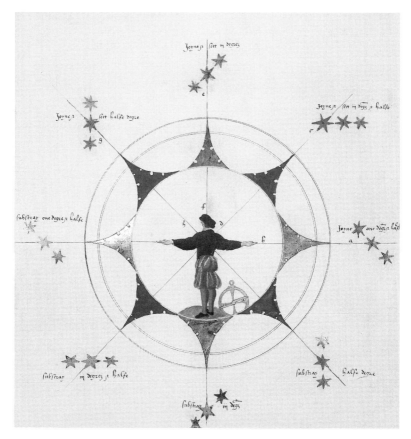

Figure 4.2 A Mariner's measure, from Jean Rotz's *Boke of Idrography*
(c. 1534).

world – that is, to draw it in and make it small by translating it into numbers derived through observational calculations performed with the mariner's astrolabe at his feet. Rotz's image, in this sense, displays the back
and forth translation of information as both the imposition of a model
upon the world and the derivation of data from it (see Chapter 1).[9]

Scale, in this sense, is closely related to practices of translation – e.g. the
translation of the heavens onto the scale of an instrument, or the obverse,

---

[9] Giving a somewhat different interpretation to Panofsky's insight that "the history of perspective may be understood with equal justice as a triumph of the distancing and objectifying sense of the real, and as a triumph of the distance-denying human struggle for
control; it is as much a consolidation and systematization of the external world, as an
extension of the domain of the self." See Erwin Panofsky, *Perspective as Symbolic Form*,
trans. C. Wood (New York, 1997), 67–68.

the disciplining of the heavens according to the measure of an arbitrary scale, or the charting of a ship's movement across the ocean. This onto-geographical embrace also moves in the opposite direction, from geodesy to the body, from the macrocosm to the microcosm, and thus to the trans-fer of the blood and guts of a body opened by a dissector's knife onto the printed page. If for Leonardo Vitruvian man was a *cosmografia del minor mondo* – a cosmography of the microcosm – his anatomies were (one of many) empirical voyages into this world. Sawday's analogy between the explorer and the anatomist thus speaks to more than a casual analogi-cal relationship, for anatomical explorations mark similar displacements as those carried out by cartographers and the ships of merchant sailors. While one conquered the body, the other conquered – scaled – the earth.

In the present chapter, I propose that we pursue a roundabout voyage across a series of scalar practices – ranging from grammar, anatomy, geodesy, cartography, and navigation – as a means of understanding where, and by what means, Jean Parmentier learned the bloody business of dissecting his men on board La Pensée; how he was able to navigate from one escale (one port) to another; and how he was able to write poetic appeals to his men that would (he hoped) persuade them to con-tinue on with their voyage. In this sense, I want to transform the question with which Parmentier's began his *Oration*, "What am I doing here?" to have him ask instead, "Where am I?" There is, I believe, a contiguity between his understanding of the anatomized bodies of his sailors as somehow falling short of – and off course in – an abstract representa-tion of the human body, and what is, essentially, a navigational question regarding the correspondence between the world of immediate sense experience and a scaled representation that allows one to see and chart the movement of ships across immense physical distance.[10] Indeed, Gens nouveaux such as Parmentier, Crignon, and Ango were defined as a class less by the possession of wealth, land, marriage, family, patronage con-nections, office, or – as Marius would aver – by the performance of great deeds, than by specific patterns of behavior, attitude, cognitive skill, and cultural disposition. Though this form of life was frequently expressed by a predisposition toward active – public – involvement in urban polit-ical, social, and cultural life (whether as designers of fêtes, members of confraternities or holders of office), it can also be discerned in a common epistemic style – an approach to, and appropriation of, the world based on acts of distanciation and translation of scale. This is most evident in the case of perspective in painting, in the cartographic techniques

---

[10] Edwin Hutchins, *Cognition in the Wild* (Cambridge, MA, 1995), 12.

developed to represent the spherical world on a flat surface, or in the metrological efforts expended to determine an accurate measure of a degree of the arc of the meridian, etc., but it can also be discerned in the writing of history; in the dissection and mapping of the human body; in the collecting practices associated with cabinets of curiosity; and in the "objectification" and "anatomization" of vernacular languages as grammars. In all these cases, epistemic theory, cognitive style, and hands-on making and doing were coterminous with acts of social and cultural distinction, that is, of distanciation. In other words, and as we shall see, new men such as Parmentier were constituted epistemically as much as they were constructed socially.[11] It is to the social and cognitive homologies between observational and inscriptive practices, and their importance in defining the form of life of the "new" men that this chapter is concerned.

Expert Territory: Words and Things

Though anatomies were an official and required part of the medical curriculum, they were not particularly common in early sixteenth-century France. Only one or two were officially performed a year, in centers of learning such as Paris, Lyon, and Montpellier. They were somewhat more common in Italy. There is evidence that postmortem examinations were not uncommon among Florentine elites in the fifteenth century.[12] Perhaps they were also a social imperative in parts of France in the sixteenth. The Paris-based Spanish physician and humanist, Andres de Laguna, for example, claimed to have dissected at least two nobles in 1535.[13] There is no evidence, however, of amateurs such as Parmentier opening up the bodies of the dead to learn their secrets. In Italy, of course, we can point to Leonardo's well-known dissections, and assume that his anatomical investigations – while perhaps not common – were

---

[11] Indeed, the distinction is redundant. See, for example, Bourdieu, *Distinction*. On debates concerned with the sociogenesis of elite identity in early modern France, see Ellery Schalk, *From Valor to Pedigree: Ideas of Nobility in France in the Sixteenth and Seventeenth Centuries* (Princeton, 1986); Jonathan Dewald, *The Formation of a Provincial Nobility: The Magistrates of the Parlement of Rouen, 1499–1610* (Princeton, 1980); John Hearsey McMillan Salmon, "Storm over the Noblesse," *Journal of Modern History* 53 (1981): 242–257; Elias, *The Civilizing Process*; and George Huppert, *Les Bourgeois Gentilshommes: An Essay on the Definition of Elites in Renaissance France* (Chicago, 1977). On widening our analytic lens to view the *gens nouveaux* as being part of a broader cultural elite than that comprised exclusively by the *noblesse de robe* see, for example, Gilbert Gadoffre, *La Révolution culturelle dans la France des humanists* (Geneva, 1997), and Wintroub, *A Savage Mirror*.
[12] See A. Carlino, *Books of the Body. Anatomical Ritual and Renaissance Learning*, trans. J. Teddeschi and A. Tedeschi (Chicago, 1999), 191–192.
[13] C. D. O'Malley, *Vesalius of Brussels, 1514–1564* (Berkeley and Los Angeles, 1964), 55.

emulated by other curious non-physicians. Could Parmentier have seen a public anatomy in Paris or Montpellier or maybe in Caen? Might he have read about them in Mondino de Luzzi's *Anathomia corporis humani* published by Antoine Blanchard, a Lyonnaise-based printer, in 1528? Did he practice on animals and learn their anatomies first? Did he watch as a postmortem examination was conducted on a friend or family member? Did he have occasion on his previous voyages to undertake similar operations? We know, for example, that just a few years later, in the winter of 1536, Jacques Cartier ordered a postmortem exam of one of his men, Philip Rougemont, aged 22, from Amboise, who had died of a mysterious and pestilential disease. Though his hope of finding a possible cure through "opening" one of his deceased men was consonant with Parmentier's stated reason for performing autopsies on board La Pensée, Cartier stood on the sidelines and left the work of dissection to an amanuensis, most likely the expedition's *barbier*, Samson Ripault.[14] Where then, did Jean Parmentier, writer of poetry, maker of maps and captain of merchant vessels, learn enough about human anatomy such that he felt qualified to identify – or at least speculate about – morbid abnormalities in his men's anatomy? We don't know, though it is fairly certain that he did not learn his anatomy at university, for as Crignon said: "*il n'a pas beaucoup hanté les escolles.*"[15] But not going to university was no barrier to curiosity. As Parmentier avers: "I have had other propitious means to have honor, as have others, in acquiring the solemn round bonnet of a doctor at the University."[16] As we saw in the previous chapter, Parmentier was explicitly referring here to his expertise and experience in navigating the sublunary world of particulars. The idea of the vita activa so eloquently expressed by Sallust through Marius (and translated by Parmentier) – that one needed to engage actively with the world and see for oneself – had widespread appeal to new men such as Parmentier. He would have tried his hand at dissection before attempting to do one on board La Pensée. In other words, his bloody experiments on his ship were surely informed by more than what he could find just in books. As Jacobus Sylvius, university professor in Paris, specialist in Galenic anatomy, and Parmentier's contemporary, said: "it is much better that you should learn the manner of *cutting by eye and touch than by reading and listening*. For reading alone never taught anyone how to sail a

---

[14] See J. P. Baxter, *A memoir of Jacques Cartier, sieur de Limoïlou, his voyages to the St. Lawrence, a bibliography and a facsimile of the manuscript of 1534, with annotations, etc.* (New York, 1906), 192.

[15] See Parmentier, *Oeuvres*, 3.

[16] Ibid., 94. See also, note 64 Chapter 3 of the present work.

ship, to lead an army, nor to compound a medicine, which is done rather by the use of one's own sight and the training of one's own hands."[17]

Though Sylvius was careful to maintain and indeed inverse the relationship between the mind and the body, words and deeds, words *were* the stock and trade of the new men – whether in commerce, law, medicine, or theology. In the first instance, bodies – and bodily practices – needed to be translated into words for moral and social reasons. Though reliance on the "real" world was emphasized, practices associated with it were translated into words as a means of establishing distance between minds that could appropriately weigh and understand, and hands getting dirty in the empirical world. Such social translations had important epistemic consequences, providing the means to create not only lasting and durable translations of experience, making them, quite literally, mobile by means of inscribed and material reproductions in books and on instruments, but also lending them a certain degree of social legitimacy insofar as they were distanciated from their entanglements in material practice.[18] In this sense, words were like Roman *imagines*, without them, individuals – and individual acts – were just that: individual, ineffable, and quickly forgotten. This was necessarily the case insofar as the replication of first-hand experience required translations into more immediately accessible – and mobile – media that could be reliably replicated and transported. Indeed, the very notion of firsthand experience becomes problematic when what one was supposed to be seeing was beyond the scale of human perception or could only be accessed by a few hardy explorers – e.g. the invisible world of nature, the order of the heavens, the path of a ship over thousands of miles, the measurement of 1 degree of the arc of a meridian, or the disposition of the internal organs of a criminal (or a sailor) splayed on a dissecting table. The work of translation in all these cases involved transformations of scale that allowed empirical phenomena (and the social practices associated with them) to be credibly distilled and transformed into inscriptions on vellum, paper, iron, and brass that were available to more than just a privileged few. In other words, a universe of particulars needed to be translated into objects and practices that could be held in the hands and seen: whether as books, maps, or instruments.[19] In this way, and with this economy of scale, the eye could, in theory, master the original without betraying it.

---

[17] Quoted in Anonymous, "Jacobus Sylvius (Jacques Dubois) 1478–1555, Preceptor of Vesalius" the *Journal of the American Medical Association*, 195:13 (1966): 1147. My emphasis.

[18] See, for example, Pamela Long, *Openness, Secrecy, Authorship: Technical Arts and the Culture of Knowledge from Antiquity to the Renaissance* (Baltimore, 2001), 176.

[19] See B. Latour, "Visualization and Cognition."

These autopic and translative practices of scale not only included cartography, navigation, and anatomy, but as I will argue presently, language and grammar.

Bodily Grammars

The first grammar of a Romance language, the *Grammatica della lingua Toscana*, has been attributed to Leon Battista Alberti, a renowned expert in scale – that is, in perspective, architecture, city planning, cartography, and practical mathematics.[20] The first Spanish grammar, *Gramática de la lengua castellana*, followed some 50 years later, written by Antonio de Nebrija, who was known to his contemporaries not only as a grammarian (as he is today), but as a polymath who wrote extensively on geodesy, the *materia medica*, botany, and cosmography. A measure of his interest in practical mathematics can perhaps be gauged by his experimental attempts to determine the value of a degree of the arc of the meridian between Burgo de Osma and Alcalá de Henares (which he found to be 20 Castilian leagues). A generation later, Fernão de Oliveira (b. 1507) produced the first vernacular grammar of Portuguese, the *Grammatica da lingoagem portuguesa*; Oliveira, like his Italian and Spanish predecessors, was not just a grammarian, he was also a cartographer-cosmographer who wrote treatises on the art of shipbuilding, naval warfare, and navigation.[21] There thus seems to be an odd contiguity between practical mathematics and grammar.

One of the first Latin French grammars was by Jacques Lefèvre d'Étaples published in 1529. D'Étaples was a religious reformer and humanist. He published and wrote widely on matters of theology and natural philosophy. He, along with his students, Josse Clichtove and Charles de Bovelles, cultivated a keen interest in mathematics, publishing or writing about Euclid, Boethius, and Nemorarius, in addition to their own mathematical work.[22] From the mid 1520s, d'Étaples had also been teaching grammar. His method was to reveal the structure of language to his students in visual form by making it accessible to their eyes as 12 charts that mapped the inner workings of language. His book, aptly named *Grammatographia ad prompte citoque discendam grammaticen,*

[20] See, for example, Samuel Edgerton, *The Renaissance Rediscovery of Linear Perspective* (New York, 1975).
[21] Fernão de Oliveira, *Livro da Fabrica das Naos* (c. 1580) and *Ars nautica* (c. 1570), manuscripts held, respectively, by the Portuguese National Library and the University of Leiden.
[22] Timothy Reiss, Knowledge, *Discovery and Imagination in Early Modern Europe: The Rise of Aesthetic Rationalism* (Cambridge, 1997), 34. Peruse, for example, Eugene Rice's *The Prefatory Epistles of Jacques Lefèvre d'Étaples* (New York and London, 1972).

tabulas tum generâles (Paris, 1529), was a translation of this pedagogical practice of mapping into prose. In the preface of d'Étaples' book, Simon de Colines directly compares this grammatical map with cosmography: "*just as, thanks to those general [universali] descriptions of the world called cosmographies,*" he said, "*anyone can quickly gain knowledge [totem oculis] of the entire world … in the same way this Grammatographia enables us to see all of grammar in a very short time.*"[23]

In 1531, one of d'Étaples' many students, a new man from Picardy by the name of Jacques Dubois, published a grammar. Dubois also went by the Latinized name Jacobus Sylvius, the physician quoted above who extolled the virtues of firsthand experience in contrast to learning through books. Dubois won renown as teacher and expert both in the *corpus Galenicum* (much of which he translated in the 1530s) and in practical anatomy. Sylvius's œuvre adds weight to the supposition that intertwining methodologies of scale were being applied to "disciplines" as diverse as grammar and anatomy. Indeed, I would go so far as to say that mapping the structure and functions of the body and mapping out the body of a language were part of the same empirical epistemo-theological practice. It was, in this sense, the explicit aim of Sylvius' grammar, as he put it, to "discover the system of the French language," to open her up and "enter into her rules."[24]

For Sylvius, language appeared on its surface to be a temporally bounded historic and geographic entity that changed according to time and place. He had an equally contextualized notion of anatomy as thoroughly embedded in historical circumstances.[25] This is why, he explains, his anatomies differed so greatly from those of his classical progenitor and model, Galen. Yet, as Kristeva points out, Sylvius was by no means a relativist; he believed, rather, that through his practice as a grammarian

---

[23] See E. Rice (ed.), *The Prefatory Epistles*, 501. For the translation here see J. Kristeva, *Language – The Unknown: An Initiation Into Linguistics* (New York, 1989), 147, emphasis added.

[24] The full quote is as follows: "Les livres de Galien sur l'usage des parties du corps humain, je les avais non seulement révisés, selon l'exemplaire grec, avec un grand soin et une ardeur particulière mais je les avais pratiquement rénovés. Alors, brisé de veilles, de soucis, de travail, j'ai recherché une matière dans laquelle je puisse réparer et renouveler les forces de mon esprit, fatiguées d'une trop longue et trop pesant étude. Aucun terrain ne me parut plus propice à ce loisir studieux que la découverte même de la langue française ainsi que son enseignment. Tandis que, sur ces deux thèmes, je suis au travail dans l'anxiété, je me rends compte que J'ai besoin d'une tension d'esprit non négligeable, tant il était labourieux de découvrir le système de la langue française et de la faire entrer dans des règles." Jacques Dubois (Sylvius), *Introduction à la langue française suivie d'une grammaire* (1531), trans. and notes, C. Demaizière (Paris, 1998), 199. My emphasis.

[25] See Roger K. French, "Natural Philosophy and Anatomy," in J. Ceard, M-M. Fontaine and J-C. Margolin, *Le corps à la Renaissance. Actes du XXXᵉ colloque de Tours, 1987* (Paris, 1990), 447–460, 452.

and an anatomist he could map out and identify not only the differences beneath one language (and one body) and another, but also discover the formal "universal" qualities that united them.[26] In other words, grammar was essentially anatomical, and vice versa – both aimed to dismember, analyze and compare particular structures and functions as a means of discovering their hidden (universal) commonalities. Kristeva explains:

> Sylvius devoted himself to transferring the categories of Latin morphology to French. To do this, he cut up utterances not only into words but also into larger segments, and looked for corresponding ones in the other language. One can deduce from this that Sylvius believed in a core of logical universals common to every language, and underlying the various construction of each language.[27]

Parmentier's project, as we have seen, was similar: particulars composed of the winds, the seas, and the motion of the stars were noted, mapped and – with the aid of instruments and charts – used to discover the God-given universals and constants that would allow him to orient his vessel on the world's seas. He perhaps had a similar map-idea in his mind against which he could compare the dissected bodies of his men, Dresaulx and the son of Pontillon.

Sylvius published his *Introduction à la langue française avec une grammaire Latino-française inspirée des auteurs hébreux, grecs et latins* with Simon de Colines's step-son and former partner, Robert Estienne, author of the *Thesaurus linguae latinae*, and brother to Sylvius' student, the translator, publisher and "anatomist," Charles Estienne.[28] In 1535, he turned to Colines to print his *Methodus sex librorum Galeni in differentiis et causis morborum et symptomatum in tabellas sex*. These six tables, as the title suggests, were modeled after d'Étaples' grammatical maps, consisting of six synoptic charts of the human body along with the associated signs and (humoral) causes of disease that could afflict them.[29]

Another student of d'Étaples, the physician Jean Fernel (b. 1497?) (see Chapter 3), not only authored the first systematic treatment of physiology since Galen,[30] and coined the word "pathology" – which he conceived as the "systematic essay on morbidity, pursued unhaltingly through the

---

[26] Compare this, for example, to Charles de Bovelles' "geography" of the French language, *Liber de differentia vulgarium linguarum et Gallici sermonis varietate* published in Paris by Robert Estienne in 1533. See Christian Schmitt, "Bovelles, Linguist," in *Actes du Colloque international tenu à Noyon, Charles de Bovelles en son cinquième centenaire: 1479–1979* (Paris, 1982), 247–263.

[27] Kristeva, *Language*, 147–148.

[28] Recall that Simon de Colines was married to the widow of Henri Estienne (the father of Robert, the lexicographer and grammarian, and Charles, the anatomist). C. Estienne, *Tres libri de disectione partium corporis human* (Paris, 1545).

[29] See C. E. Kellett, "Sylvius and The Reform of Anatomy," *Medical History*, 5:2 (1961).

[30] Fernel, *The physiologia*, 4.

body, organ by organ"[31] – but was also an important instrument maker, mathematician, and cosmographer. Fernel did not write grammars of languages but rather grammars of the body, using dissection to bring all its parts, one by one, into view;[32] in addition, he worked tirelessly in the realm of practical mathematics in the area we now call geodesy. Thus, well before he published his *De naturali parte medicinae* (1542), he published (with Colines) works of practical mathematics, such as the *Monalosphaerium* (1526), *De proportionibus* (1528) and *Cosmotheoria* (1528).

In the *Cosmotheoria*, for example, Fernel calculated the measure of a degree of the arc of the meridian between Paris and Amiens. He did this not by reading ancient texts, but as Marius would have it, with his eyes and hands; which is to say – much like Parmentier in navigating across the earth's surface, or Alberti measuring the bodies of men and statues with a finitorium, or Nebrija undertaking a similar metrological investigation in Spain – he did this experimentally,[33] taking the height of the sun in Paris (with a cross staff or astrolabe), and, after measuring the circumference of a carriage wheel, counting the number of times it turned as he traveled north toward Amiens (17,024 times), and then once again, taking the height of the sun, all the while keeping time with a spring-built clock, his "trusted" horarium, so that he could compare local time in Paris with that at his destination in Amiens.[34]

Though Fernel won fame as a physician, some of his most important accomplishments were in the domains of practical mathematics. He clearly saw his work as furthering the progress of geography and transoceanic navigation:

Everyone knows that it is through navigational skill, rather than because of eagerness for new discovery, that ships have traversed the Ocean, discovered the Isles, *opened up* the innermost recesses of the Indies, *revealed* to our people (it did them much good) a great part of a continent to the West, hence named the "New World," of which the ancients were unaware. This [skill], like all the astronomy that was not sufficiently visible to Plato, Aristotle and the older philosophers, was greatly built up and illuminated later by Ptolemy; but if he returned to life now, he would not recognise Geography, so much has a New World been *brought into view* by the sea travel of this epoch. I do not claim that we have helped sea travel along; but through observation of the equinoctial hours, we have certainly

---

[31] C. Sherrington, *The Endeavour of Jean Fernel* (Cambridge, 1946), 101, quoted in Ibid., 2.
[32] Ibid.
[33] Taken in its French sense, as experience.
[34] See the introduction to *Jean Fernel's The Hidden Causes of Things*, trans. John M. Forrester with introduction and annotations by J. Forrester and J. Henry (Leiden and Boston, 2005), 10. His calculations were not improved upon until the next century.

worked out the basis for recognising what Geographers call the longitude, wherever one may be in the globe.[35]

This intertwining of navigation with practical mathematics by Fernel, one of the preeminent physicians in France, speaks to a set of cultural dispositions and attitudes shared by men such as Sylvius, and the man who taught them both, Jacques Lefèvre d'Étaples;[36] it can also be extended to their students, such as the author of *De humani corporis fabrica libri septem* (1543), Andreas Vesalius who began his medical studies in Paris with Sylvius and Fernel. Vesalius left France to continue his studies at the University Louvain, where he worked with the renowned mathematical instrument maker Gemma Frisius. It was with Frisius that he first absconded with the corpses of executed criminals to dissect. For his part, Frisius, in addition to being a cartographer, mathematician, and instrument maker was Regius Professor of Medicine at the University of Louvain.

Frisius was famous throughout Europe not only for the accuracy of his instruments, which were praised by Tycho Brahe, among others, but for having been the first to put forward the method for mapping distance through triangulation, and for proposing that one could determine longitude by using an accurate clock and an astrolabe.[37] He is perhaps best known as a popularizer of Peter Apian's influential *Cosmographicus liber* (1524) through several best-selling re-editions.[38] Along with his student, Gerardus Mercator, he also made globes and maps.[39] Mercator, in turn, is best known for the eponymous Mercator projection, which finally realized the promise of the map by allowing navigators to follow their progression simultaneously in abstract and "real" space "without" distortion; his cylindrical projection allowed for the maintenance of scale and the disciplining of space through complicated mathematical formulae such that a rhumb line – a course – could be followed by the mind

---

[35] Fernel, *The Hidden Causes of Things*, 109. My emphasis.

[36] D'Étaples was a prolific translator of books of natural philosophy, astronomy, cosmology, mathematics, theology, and rhetoric.

[37] See, for example, H. Cook, "The New Philosophy in the Low Countries," in R. Porter and M. Teich (eds.), *The Scientific Revolution in National Context* (Cambridge, 1992), 118.

[38] Steven Vanden Broecke, "The Use of Visual Media in Renaissance Cosmography: the Cosmography of Peter Apian and Gemma Frisius," *Paedagogica historica*, 36:1 (2000): 131–150.

[39] It is interesting to note that according to Frisius, "the utility, the enjoyment and the pleasure of the mounted globe, which is composed with such skill, are hard to believe if one has not tasted the sweetness of the experience. For, certainly, this is the only one of all instruments whose frequent usage delights astronomers, *leads geographers, confirms historians, enriches and improves legists [les legistes], is admired by grammarians, guides pilots*, in short, aside from its beauty, its form is indescribably useful and necessary for everyone," quoted by Jerry Brotton, *Trading Territories: Mapping the Early Modern World* (London, 1997), 20, my emphasis.

as well as by ships as a constant bearing. In addition to mapping courses across the physical world, Mercator understood the art of cartography, as Parmentier understood his voyage, to be a spiritual exploration: a vehicle for contemplating "the magnificence of God's creation" which he saw as being directly related, perhaps not surprisingly, to the "divine perfection manifested by the human body."[40]

Another of Vesalius's teachers was Johann Winter von Andernach (b. 1487). Winter was well known as an anatomist, but he was also a respected Greek grammarian and lexicographer. In addition to publishing widely in the medical arts, such as his translation of Galen's *De Anatomicis Administrationibus* (1531), or his 1536 dissection manual, *Institutiones anatomicae*, he also wrote and published a Greek grammar in 1527. He praised Vesalius as "a youth of great promise," noting that he possessed "a remarkable knowledge of medicine and of Greek and Latin, and great dexterity in dissection."[41] Another of the able dissectors he named was Michael Servetus, who was the first anatomist to describe pulmonary circulation. Like Vesalius, Servetus studied medicine with Sylvius, Fernel, and Winter,[42] but he was equally well known for his skills as a translator, mathematician, astrologer, geographer, and theologian. Among the many works that he published was the first French edition of Ptolemy's *Geography* (Lyon, 1535).

When Pierre Crignon published Jean Parmentier's *Exhortation* in 1532, he turned to Winter's compatriot and friend, Gérard Morrhy (Morrii, Morrhe, Morrhius) based in Paris. In his brief career, Morrhy published an impressive array of books, including Galen's commentary on Hippocrates' *de Natura Humana*, the *Rhetoric* and the *Poetics* of Aristotle and a Greek-Latin lexicon that he himself authored. Morrhy also published the 1532 edition of the *Protomathesis* by the cartographer, translator, judicial astrologer, instrument maker, and Regius Professor of Mathematics at the *Collège Royal* – the *Collège des Trois Langues* – Oronce Fine. Fine studied with Jacques Lefèvre d'Étaples and published a great many of his works with Simon de Colines. He worked closely with Bovelles, and utilized Fernel's measurements of latitude in his map of France, the *Nova Totius Galliae Descriptio*, published with Colines in 1525.[43] Fine's world maps of 1531 and 1534, also printed by Colines,

[40] Catrina Albano, "Visible Bodies: Cartography and Anatomy," 89–106, at 89 in A. Gordon and B. Klein (eds.), *Literature, Mapping, and the Politics of Space in Early Modern Britain* (Cambridge, 2001).

[41] Quoted in Max Fisch, "Vesalius and His Book," *Bulletin of the Medical Library Association* 8 (1943): 208–221, at 210.

[42] J. Trueta, "The Contribution Of Michael Servetus to the Scientific Development of the Renaissance," *The British Medical Journal* 2 (1954): 507–510, at 509.

[43] F. de Dainville, "How Did Oronce Fine Draw His Large Map of France?" *Imago Mundi*, 24 (1970): 49–55.

included recent geographical intelligence from voyages sponsored by
Ango, utilizing, for example, information supplied by Verrazano's voyage
of 1524 in charting the eastern coast of North America (see Chapter 6).
Among his most interesting and influential maps were those made in
the shape of a human heart, which – according to at least one sixteenth-
century commentator – were called cordiform precisely because they
truly displayed "the image of the heart of living beings."[44] Given the
shape of Fine's world anatomy, it is perhaps not surprising that in addition to being a mathematician and cartographer, Fine was also trained –
like his father and paternal grandfather before him – as a physician.

We can more firmly connect Parmentier to these circles by tracing
another parallel network that intersects with the publisher, Jean Pierre
de Tours, and the printer, Simon du Boys or Dubois, of Parmentier's
1528 translation of the *Catiline Conspiracy*. In the 1520s Simon Dubois
printed books for a living on the street of the traitor, *la rue Judas*, in Paris.
He published a number of works by both Erasmus and Luther in 1525.
A little later he worked with Geoffroy Tory on his ground-breaking *Book
of Hours* (1527). Tory began his professional life under the tutelage of
Simon de Colines. He closely studied the works of Leon Battista Alberti,
Luca Pacioli, Albrecht Dürer, and Charles de Bovelles, being particularly
interested in researching the relationship between mathematics, the proportions and scale of the human body, and the design and mark of letters.
Tory's *Book of Hours* was a typographical landmark; it was among the first
printed books to free writing from the human hand, thereby giving material form to an increasingly systematized perspectival mapping of language on the printed page. As Tom Conley has put this, reading became
with Tory "tantamount to a mensuration of words, that is, of imagining
them as equally sized units… that are seen in a serial and discrete config-
uration."[45] Tory followed his *Book of Hours* with the *Champ fleury* (1529),
a book of grammar and eloquence of the French language as mediated
through the microcosm of the mathematized human body as letter. Tory's
work provided a kind of language map, a cartographic text, that would
enable readers not only to navigate the pages of a book, but to voyage into
the history of the French language so as to discern its fixed constancy –
like the unchanging golden ratio revealed by the structure of the human
body – amongst an ever-changing semantic topography (see Figure
4.4). *Champ fleury's* subtitle is particularly interesting given our present

---

[44] Jacques Severt, writing in 1598, quoted in Robert W. Karrow, Jr., *Mapmakers of the
Sixteenth Century and Their Maps: Bio-Bibliographies of the Cartographers of Abraham
Ortelius, 1570* (Chicago, 1993), 171.
[45] Tom Conley, *The Self-Made Map: Cartographic Writing in Early Modern France*
(Minneapolis and London, 1996), 67.

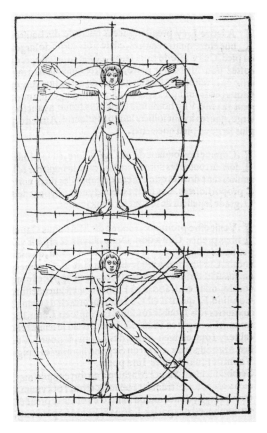

Figure 4.3 Geoffroy Tory, the Design of Letters scaled to human proportion, *Champ fleury* (Paris, 1529).

concerns: *The Art and Science of the Proportion of the Attic or Ancient Roman Letters, According to the Human Body and Face*. Indeed, there was more than a little anatomy in Tory's mapping of the letter.

The iconic image from the *Champ fleury*, which so closely resembles Leonardo da Vinci's Vitruvian Man, was drawn by Jean Perréal, a French artist, sculptor, and poet, Valet de Chambre and painter for Charles VIII, Louis XII, and François I, who knew and corresponded with Leonardo (see Figure 4.3).[46] Perréal participated

---

[46] Perréal was one of Geoffroy Tory's teachers; he was responsible for at least some of the illustrations, especially the woodcuts that demonstrate the proportions of letters based on the human form, included in the *Champ fleury*. See Auguste Bernard, *Geofroy (sic) Tory, peintre et graveur, premier imprimeur royal, reformateur...* (Aubry, 1857); and William Ivins, "Geoffroy Tory," *The Metropolitan Museum of Art Bulletin*, 15:4 (1920):

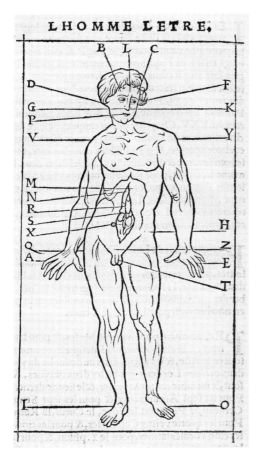

Figure 4.4 Geoffroy Tory on human anatomy and the letter, the muses, the cardinal virtues, the liberal arts and the graces and their correspondences with the vital channels and noble organs of the body, *Champ fleury* (Paris, 1529).

in the Puy de palinod of Rouen in the 1520s.[47] His poetry was included, along with that of Jean Parmentier, Jacques Le Lieur, and Jean and Clement Marot, in a manuscript commissioned by Jehan Lalement the younger from Bourges (1536). Lalement was an important patron of arts and letters, who commissioned a number of manuscripts

79–86. One might speculate that Perréal and the Juste were involved with the design and ornamentation of Ango's dockside house, La Pensée, which was reputed to be decorated with sculptured bas-reliefs à la Florentine.

[47] On Perréal's participation in the Puy see Gérard Gros, 207–208.

in the 1520s and 1530s.[48] From a family of printers, he was an early backer of his compatriot from Bourges, Geoffroy Tory. Tory worked in the royal library at Blois in the mid 1520s when it was being directed by the humanist founder of the Collège de France, Guillaume Budé, a maternal uncle of Jacques Le Lieur, one of the Puy de palinod's principle backers in the first half of the sixteenth century.[49] Of the 43 authors, and 149 poems, included in Lalement's manuscript, 23 poems were by Le Lieur, many more than by any of the other contributors (there were, for example, five by Parmentier), perhaps indicating the ties between the man who commissioned the manuscript and the Rouennais poet.[50] Two particularly fine miniatures included in a presentation manuscript of palinodic poetry commissioned by Le Lieur, BN Ms. Fr. 379 have been attributed to Perréal, those illustrating the prize-winning poems of 1518 and 1519: Le Lieur's *Sans vice aulcun toute belle conceue* and Crignon's *Pourpre excellent pour vestir le grant roy*. Perréal's possible authorship of these, and other, miniatures for Le Lieur's manuscript is given further credibility by the lengthy poetic exchange between the two men conducted in late 1527–1528.[51]

The same year as he worked on Tory's *Book of Hours* (1527), Dubois also published Alexandri Benedicti's Galenic anatomy, *Anatomice, siue, Historia corporis humani*; and, with the bookseller-publisher, Galliot du Pré, a work entitled *chantz royaulx oraisons et aultres petitz traictez* by the famous Norman poet, Guillaume Crétin.[52] Here another connection with Parmentier appears, as Crétin was Parmentier's contemporary and competed with him in Rouen's Puy de palinod in 1516 and 1520. The connections do not stop here, however. The years 1528 and 1529 were especially busy for Dubois; working with Chrétien Wechel, he produced a collection of medical texts (1528) that included Galen, Johann Winter, and Antonio Benivieni's *Libellus de abditis nonnullis ac mirandis morborum & santitationum causis*. Benivieni's text was among the first to recount details of autopsies and postmortem dissections as a means to discover the hidden causes of disease. Surely this must have

[48] See, for example, Myra Dickman Orth, "Two Books of Hours for Jean Lallemant Le Jeune," *The Journal of the Walters Art Gallery* 38 (1980): 70–93.

[49] Jacques Lèfevre d'Étaples succeeded Budé in directing the library in 1526.

[50] Le Lieur won first prize in 1518 for his chant royal, *Sans vice aulcun toute belle conceue*, with Jean Parmentier taking second prize for his *La forte nef toute plaine de grace*.

[51] E. Picot, *Notice sur Jacques Le Lieur echevin de Rouen sur ses heures manuscrites* (Rouen, 1913).

[52] Galliot du Pré worked with, amongst others, Simon de Colines and Pierre Vidoue. See Arthur Tilly, *Studies in the French Renaissance* (Cambridge, 1922), 168–218.

been one of the textual resources at Parmentier's disposal.[53] He also worked with Wechel on tracts by Luther, Fernel, and d'Étaples. Dubois also worked with Pierre Vidoue in 1528 on the physician, astrologer, and mapmaker Laurent Fries' *Syderal Diuinement ou Prognostique.* Significantly, in the present context, Vidoue also introduced and edited a collection of palinodic verse by *"scientifiques personnaiges"* from the Puy de palinod of Rouen (1525) that included poems by both Crétin and Pierre Crignon.[54]

Dubois, of course, also worked with Jean Pierre de Tour, with whom he produced two books: Oronce Fine's *La Théorique des cielz, mouvemens et termes practiques des sept planets* and Jean Parmentier's translation of Sallust's *Catiline Conspiracy.* There is little information available on Jean Pierre de Tours other than after working on these two books with Dubois he began working with Gérard Morrhy. It is likely that after Dubois was forced to flee Paris because of his religious views, Crignon turned to his partner in the production of Parmentier's translation, Jean Pierre de Tours, who then brought him to Morrhy to publish the *Oration.*[55]

From this brief examination of the intellectual reach of some of Parmentier's more learned contemporaries, and the network of book publishing which entangled him with men such as Lefévre d'Étaples, Simon de Colines, Jacques Dubois, Jean Fernel, Gemma Frisius, Andreas Vesalius, Gerardus Mercator, Geoffroy Tory, and Oronce Fine, it is clear that his interest in practical mathematics (cartography, instrument making, geodesy, navigation), eloquence, poetry and grammar, and human anatomy were not entirely idiosyncratic and exceptional, but were part of a cohesive cultural world that centered on acts of distanciation, scale and taking the measure of things. Of course, these engagements were in some sense underwritten by the close connections between medicine and astrology, with mathematics acting as a mediator between the universal aspects of the cosmos and the microcosm of the body. However, the connection between anatomy, practical mathematics, and grammar takes us in a slightly different direction. In this sense, I am not making an argument from analogy, arguing that anatomy was like grammar, which was like cartography, which

---

[53] Wechel also produced a number of other editions in 1528 that included Benivieni's text.

[54] Pierre Vidoue (ed.), *Palinods, chants royaulx, ballades, rondeaulx et epigrammes à l'honneur de l'immaculée Conception de la toute belle mère de Dieu Marie / composés par scientifiques personnaiges des clairés par la table cydedans contenue* (Paris and Caen, 1545).

[55] Parmentier's work is marked only by Morrhy's device and his adage: "nocet empta dolore voluptas." See Philippe Renouard, *Imprimeurs parisiens, libraires, fondeurs de caractères et correcteurs d'imprimerie*, Google eBook (Paris, 1898), 299.

was like anatomy, which was like celestial navigation, but rather, I am attempting to map out a field of related and overlapping observational and inscriptive practices that demarcated the social, epistemological, and theological terrain inhabited by France's New Men, men such as the map and instrument maker, poet and translator, ship-captain and "amateur" anatomist, Jean Parmentier.

Anatomy of the Heart

The heart-shaped map drawn by Oronce Fine in Figure 4.5 can, I believe, take us a step further into this cultural world. Maps were once objects of religious devotion; they were displayed on altars and charted sacred pilgrimages, the geography of holy sites, and the (unreachable) location of the terrestrial paradise.[56] Fine's maps followed in this tradition. Indeed, to understand the significance of Fine's cordiform maps, not only is it important to understand the significance of the heart in early modern French medical thought, but also its role as an emblem that could move from the pages of a book, a cordiform map, or a theatrical display, to scale theological and socio-political notions of order and authority.

According to Fernel the heart is at "the origin of all faculties, because it is the grandest part. It is as if a king, the grandest in all his realm, were to appropriate the powers of all his subjects to himself."[57] The heart was an absolute ruler that had no need for councilors and brooked no freedom among those it ruled. But, though the heart was infused with vital spirit, the very force and substance of life, it was also, according to Parmentier's contemporaries, the seat of the passions. In this sense, its rule was subject to the vagaries of ambition, avarice, anger, and fear. One could locate everything that Sallust saw as symptomatic of Rome's moral decline and everything Parmentier feared from his crew in the space and functioning of the heart. In ancient Rome as in early modern France, with Sallust as with Parmentier (and, indeed, his patron, Ango), it was the brain's task to moderate the passions and bring them into the service of reason. Recall, for example, that both Parmentier and Ango directed their respective worlds – their empires – from a ship and house that were both called La Pensée.

In a similar way, Fine's cordiform map was an attempt to contain and subsume the accidents and particularities of the material world – its

---

[56] See Flint, *The Imaginative Landscape*, 6–7.
[57] Fernel, *The Physiologia*, 379. The surgeon, Henri de Mondeville said much the same thing at the beginning of the fourteenth century in comparing the heart to "un roi au milieu de son royaume." Quoted in M. Gaude-Ferragu, "Tombeaux et funérailles de coeur en France à la fin du moyen âge," in *Il cuore/The Heart* (Florence, 2003), 243.

Figure 4.5 Oronce Fine, Cordiform Projection (1531).

passions – and bring them toward the universal through their ordered mathematical disposition.[58] Like Parmentier's crew, the material world in Fine's map was to be governed by reason – by thought. Robert Karrow has remarked that though Fine's cordiform projection was not a cartographically practical solution to the problem of representing the globe on a two-dimensional surface, it was "an elegant way of demonstrating the application of transcendent mathematical principles to terrestrial things."[59] We should be careful however not to credit mathematics

[58] According to Mary Carruthers, in the post-Aristotelian tradition, it was the heart that received "externally-derived impressions," while the brain was the site of their reception and storage. Though physiologically memory was located in the brain, she explains, the heart maintained a metaphorical association with memory, as exemplified by the Latin verb *recordari*, meaning to "recollect." See Carruthers, *The Book of Memory. A Study of Memory in Medieval Culture* (Cambridge, UK, 1990), 48–49.

[59] Cited by Giorgio Mangani, "Abraham Ortelius and the Hermetic Meaning of the Cordiform Projection," *Imago Mundi* 50 (1998): 59–83, at 65. See Karrow, *Mapmakers*, 179.

with a status it did not yet have in the early sixteenth century. Practical mathematics, despite the pronouncements of mathematical practitioners such as Fine, was an inferior "material" exercise not a discipline, let alone a "transcendent" one, like philosophy or theology. Measuring the world, like measuring the motions of the planets and the stars, was not a particularly laudatory endeavor. Mathematics, rather, was considered propaedeutic to the study of medicine insofar as it was necessarily part of astrological and astronomical learning, both of which were considered crucial to diagnostics and therapeutics. Its practitioners, with a few notable exceptions, inhabited a lesser rung on the social hierarchy. Jean Fernel, for example, was forced to give up mathematics for medicine precisely for these reasons.[60] As Guillaume Plancy, Fernel's student and biographer wrote (imagining Fernel's father-in-law presenting his case to his future son-in-law as to why he ought to abandon mathematics for medicine):

> Mathematics made no contribution to the public weal. Apart from a modicum of arithmetic and geometry it touched society little or not at all. On the other hand when we turn our gaze and thought to medicine we find it a science occupied either with sublime enquiry into Nature or with deeds of beneficence and utility. It is of right the worthiest of all the arts. Mathematics offers no comparison with it.[61]

One of the strategies embraced to elevate the status of mathematics, as we have seen, was to link it to the glories of imperialism through the promise of better navigational techniques, geographical intelligence, and, of course, profit. In this sense, Fine's maps were very much concerned with "deeds of beneficence and utility." By "flattening" heaven and earth so that they could be projected and inscribed on brass, vellum, and paper, Fine made it theoretically possible for the world to be traversed not only physically by ships and men (a purpose for which his maps were never employed), but by the eyes of merchants, sailors, and kings without them ever having to leave the shores of France. This "utilitarian" marketing role could not bear the entire weight of lifting the status of mathematics (practical or otherwise) beyond its disciplinary limbo as a stepping-stone to higher things, e.g. medicine. If utility was to become *scientia*, that is, true and laudable knowledge, it would also need to become universal in its own right. Fine therefore relied upon his vocation as a physician to direct his expertise as a mathematician toward what Plancy called the "sublime enquiry into Nature." Fine's map, in this sense, attempted to

---

[60] John Henry, "*Mathematics Made No Contribution to the Public Weal*: Why Jean Fernel (1497–1558) Became a Physician," *Centaurus* 53 (2011): 193–220.

[61] Quoted in ibid., 208.

conjoin practical reason to the human heart, and thus to natural philosophy and theology in the same way that Crignon tied the astrolabe to the composition and ritual recitation of Marian verse. Fine's mathematically made heart, constructed out of the "words and observations of many hydrographers," encompassed the earth from the perspective of the heavens, it contained and constrained the seat of the passions, and thus demonstrated the mind's power (La Pensée) – and the power of mathematics – over the exigencies of the sublunary world.

It is interesting to note, in this regard, that Fine was not only a mathematician, cartographer, astrologer, and physician, but like d'Étaples, Sylvius, Fernel, and Winter, a prolific translator and editor. One of the many texts he helped publish was by the expert in "Christian" cabala, the converted Jew, Agostino Ricci called *De motu octavae sphaerae*, published by Coline in 1521. *De motu*, to which Fine wrote the preface, dealt with the "movement of the eighth sphere in a philosophical and mathematical manner, together with the teachings of the Platonist and ancient magic (which the Hebrews call Kabbalah)."[62] Fine, through Ricci, was connected to Agrippa and, indeed, to much larger networks of neo-platonic and hermeticist astrologers.[63] It was these – for some, shady – connections, along with his own astrological practice, that landed him in jail for several years in the 1520s.

The heart in the neo-platonic thought of men such as Fernel and Agrippa,[64] was a microcosm of the operation of the *spiritus mundi*; this was an old idea, eloquently recounted in the influential work, *Cosmographia*, by the twelfth-century Neo-Platonist, Bernardus Silvestris.[65] The heart, he said, is "the animating spark of the body, nurse of its life, the creative principle and harmonizing bond of the senses; the central link in the human structure, the terminus of the veins, roots of the nerves, and controller of the arteries, mainstay of our nature, king, governor, creator."[66] The heart embraced and incorporated – through its centrality in and rule over a distributive nexus of vital spirit – the entire body in much the same

---

[62] Nicholas Goodrick-Clarke, *The Western Esoteric Traditions: A Historical Introduction* (Oxford, 2008), 58.

[63] Such as Ricci's brother, Paulus Ricius, who was a well-known popularizer of the cabbala. Also, recall that d'Étaples spent time with both Pico and Ficino, and that Symphorien Champier (the neo-Platonist physician from Lyon) was also closely related to their circle.

[64] See, for example, Giancarlo Zanier, "Platonic Trends in Renaissance Medicine," *Journal of the History of Ideas* 48 (1987): 509–519.

[65] Bernardus Silvestris, *The Cosmographia of Bernardus Silvestris*, trans. Winthrop Wetherbee. This work profoundly influenced Cusa; d'Étaples was responsible for publishing Cusa's collected works in 1514.

[66] Ibid., 125.

way that Fine's map embraced and incorporated the entire body of the world, or as the king (or, indeed, any "appropriately" skilled observer) was to possess – and rule over – the world presented to him by a map. Put somewhat differently, the sublunary world of base and contingent particularities was vouchsafed by a continuous and dynamic exchange of properties – a translation – with the archetypal forms as communicated to the heavens through the *anima mundi*; the earth through the *spiritus mundi*; the body, through the actions of the heart; and the *royaume de France* (and beyond – to empire!) through the actions of the king and his agents – i.e. men such as Jean Ango, Jean Parmentier, and Pierre Crignon.[67]

Fine's cordiform maps, similarly acted to animate and assemble the disparate parts of the material world and unite them into one body, one system – a world, an empire – made through mathematics, cartographic skill, and the reports (and daring skills) of sailors, explorers, and merchants. It is interesting to note that Agostinio Ricci, in addition to being a physician and a famous cabbalist, was also one of only two known students of Abraham Zacuto (or Zacut).[68] Zacuto was a Jewish mystic, physician, and astronomer credited by many with establishing the declination and calendric tables that were subsequently translated into step-by-step guides – like the *Regimento do Estrolabio* – that enabled mariners to estimate latitude through meridian observations of the sun and navigate the open seas.[69] The material world was, in this sense, a mirror of the heavens in which each point and every location could be identified – using a mariner's astrolabe and the appropriate declination tables – by corresponding celestial coordinates. Fine's map was a mathematical translation of this heavenly order onto a projective plane

[67] This recalls the late medieval and early modern practice of interring the king's heart (and entrails) separately from his body as relics that made mobile and extended the presence of his immortal *dignitas* across geographical (national/imperial) space. Interestingly, this attempt to remap France through royal relics was vehemently opposed by the church; As E. A. R. Brown notes, "Boniface VIII condemned this practice as savage and inhuman." See her "The Ceremonial of Royal Succession in Capetian France: The Funeral of Philip V," *Speculum* 55 (1980): 266–293.

[68] See Nicholas Goodrick-Clarke, *The Western Esoteric Traditions* (Oxford, 2008), 57–58.

[69] See Law, "On the Methods of Long Distance Control"; see also Mariano Gomez-Aranda, "The Contribution of the Jews of Spain to the Transmission of Science in the Middle Ages," *European Review* 16 (2008): 169–181; Ella M. J. Campbell, "Discovery and the Technical Setting:1420–1520" *Terrae Incognitae* 8 (1976): 11–18; J. D. North, "Essay Review: Some Jewish Contributions to Iberian Astronomy" *Aleph: Historical Studies in Science and Judaism* 2 (2002): 271–278; W. G. L. Randles, "The Emergence of Nautical Astronomy in Portugal in the XVth century," *Journal of Navigation* 51 (1998): 46–57; José Chabás and Bernard R. Goldstein, *Astronomy in the Iberian Peninsula: Abraham Zacut and the Transition from Manuscript to Print*, Transactions of the American Philosophical Society, New Series, 90:2 (2000), which challenges much of the scholarship on Zacut's

consisting of the contingencies of world, and then placed, arranged, and mathematically "corrected" using the observations of many mariners and hydrographers. Heaven thus came down to earth in the form of a heart-shaped map that could be held in the hands and grasped by the eyes. But observers here not only gazed out upon the world through Fine's map, they were also drawn inward to embrace the microcosm of the living heart beating in their chests.

Sursum corda: Lift up our Hearts Together

The relationship between microcosm and macrocosm was not considered a metaphor in the sixteenth century, it was real, and indeed, by definition, metonymic (though perhaps without the vanity of more modern notions of exactitude)[70] – God was everywhere present in this world, just as he was everywhere present in the hearts of his children. One could say the same for the *spiritus mundi*, or the Church. This integration of the material and the heavenly, the individual and the group, the particular and the universal, constituted an ideal of perfect concord (*concordia*: *con* – together + *cord* – heart) that defined the social, religious, and epistemic aspirations (and, of course, the anxieties) of men such as Parmentier, Crignon, Fernal, Sylvius, and Fine. When Parmentier sought council on board La Pensée, it was to the "Ecclesiastique" that he turned. Perhaps one of the phrases that stuck in his mind was chapter 3 verse 2 – "He hath made every thing beautiful in his time: also he hath set the world in their heart, so that no man can find out the work that God maketh from the beginning to the end."

In the fifteenth and sixteenth centuries, it was widely held that God's presence on earth could be located in the individual believer's heart. Community arose through acts of collective reading; as the anaphora to the liturgy of the Eucharist says: "May the Lord be in your heart and on your lips that you may worthily proclaim his Gospel."[71] This was understood as an inscriptive process – the heart was a book upon which God's words were written so as to be read, spoken and proclaimed to the world. The Word, as written on the heart was to circulate from the interiority of the believer's soul out into Christian *communitas*. Paracelsus' injunction that the physician should tread the pages of God's creation finds an anatomical parallel in the microcosm of the heart, whether translated into the material world made by God on Fine's map, or as His written Word, found for

---

role in (though not his influence on) the development of navigational techniques in the age of discovery.

[70] See J. L. Borges's "On the Exactitude of Science," in *Collected fictions*, A. Hurley trans (New York, 1998), 325.

[71] See Eric Jager, *Book of the Heart* (Chicago, 2000), 122–123.

example, in Nicolas Blairié's fifteenth-century cordiform Book of Hours (see Figure 4.6). It was by exploring the heart's excorporated translation (as a book of prayer or a map) that the observer could voyage into his own heart and thereby approach – and be assimilated into – an invisible world beyond the scale of human cognition, whether of geography and empire, or of God and Christian community. In both cases, this was not a question of the individual heart, but of a collectivity, a community – the heart's prayer was a form of circulation that carried individual devotion into the sacred body of Christendom, or as in the case of Fine's map, into universal Christian empire. God's word, whether written upon the material of the world in the particularities of geography, or as words inscribed upon the individual's heart, demarcated a nexus of distribution and translation to be carried out – and replicated – by the *spiritus mundi*, the Holy Spirit, and, of course, by mapmakers and good Christian navigators circulating around the globe. It was this translation, extending through economies of scale from the microcosm of the heart into the macrocosm of an "imagined community," to use Benedict Anderson's phrase, that constituted the ideological animus of such devices as Crignon's astrolabe, Fine's maps, and Blairié's *Book of Hours*. The power of the map, like the power of a ship or an astrolabe, or indeed, of a book of prayer, rested in its ability to translate – through changes of medium and scale – not only the observed, but also the observer.

Through mathematical skill combined with the observations of many mariners and hydrographers the world had become – by the end of the fifteenth century – perceptible as a geographic entity defined by mathematical space for the first time since Ptolemy. At the very same time as the world was undergoing this dramatic transformation, so too was the observer. By looking inwards, whether at the universal forms beneath the vagaries of human anatomy, or to the heart upon which God's words were inscribed, the individual was to be transported and transformed – translated – into a Christian subject, a citizen, a microcosm of this revealed world. In other words, Fine's and Blairié's cordiform instruments were translating machines of scale – they made it possible to chart journeys from the individual to the communal, and from the particular to the universal, and then back again. In this regard, it is worth noting that both instruments also had specifically social functions that were bound to their practical, epistemic, and spiritual uses. Simply put, they were prestige items insofar as they were successful in translating the world of particulars into the language of the universal – i.e. as a book encompassing God's Word or a map collecting his Works; in either case, their possession (as with the possession of a relic) mapped out – through networks of gift giving, display, and patronage – social topographies of status, authority,

Figure 4.6 Cordiform Book of Hours, à l'usage d'Amiens (fifteenth century).

and prestige.[72] What better way to do this – in a world characterized by the perception of moral decline, avarice, duplicity, and dissimulation, by social climbers and *parvenus*, by ambitious *gens nouveaux* and rapacious nobles – than by making the heart visible for all the world to see.

Concordia: The Heart and the Thought of Man

Concordia as a political and social ideal had widespread currency in Rome of the last century before the common era.[73] Sallust, for example, in the opening pages of the *Catiline Conspiracy*, reflected nostalgically on the concord that characterized Rome in its heyday.[74] "The city of Rome" he said,

---

[72] See, of course, Peter Brown, *Cult of the Saints*.
[73] Thomas Wiedemann, "Sallust's 'Jugurtha': Concord, Discord, and the Digressions," *Greece & Rome* 40 (1993): 48–57; see also R. Syme, *Sallust* (Berkeley and Los Angeles, 1964), 254. Gordon P. Kellym, *A History of Exile in the Roman Republic* (Cambridge and New York, 2006), 10; and John Alexander Lobur, *Consensus, Concordia, and the Formation of Roman Imperial Ideology* (New York, 2008).
[74] See, for example, Robert Brown, "Livy's Sabine Women and the Ideal of Concordia," *Transactions of the American Philological Association*, 125 (1995): 291–319. According to

was at the outset founded and inhabited by Trojans, who were wandering about in exile… and had no fixed abode; they were joined by the Aborigines, a rustic folk, without laws or government, free and unrestrained. After these two peoples, different in race, unlike in speech and mode of life, were united within the same walls, they were merged into one with incredible facility, so quickly did harmony (*concordia*) change a heterogeneous and roving band into a commonwealth.[75]

But as we have seen, for Sallust at least, this Rome had disappeared. While concord could make the smallest states great, he argued, "discord undermines even the mightiest empires."[76] In sixteenth-century France, as in ancient Rome, discord was understood to be indicative of moral, social, and political decline. As we saw previously, conflict between the New Men and the old nobles, between both these groups and the peasantry, and among all of them together, was thought to be responsible for the world's descent into madness and chaos. Just as the heart acted as the animating center of the corporeal body, drawing it together as a unified whole, so too, as a political emblem, the heart acted symbolically to connect and integrate – to traverse scale and to harmonize – the disparate parts of the body politic into a community. Lecoq lists a number of examples of this, e.g. the 1431 entrée royale of Henry VI into Paris, where the king was met by the "ship of state" crewed by the different Parisian estates who presented him with hearts symbolizing municipal-monarchical concord;[77] Charles VIII's entry into Troyes in 1486 where the king was given a heart by three young virgins in a fountain representing the virtues of the city's different estates. "Three we are, in a heart united," they sang.[78] Similarly, for her entry into Paris (1504), Anne de Bretagne was presented with a huge heart painted red and blue (the colors of Paris) by the city's three estates.[79] We could add a more imperial version of concordia with Henri II's entry into Rouen (1550), where the king was represented with a vine growing from his heart, and with the peoples of many nations kneeling before and around him drinking his blood – the liquor of sweet and *amyable confederation et obeissance*.[80]

Anne Marie Le Coq, "in Roman antiquity Concordia was the object of cult which was both public and private. It symbolized not only the affection between members of the same race or the same family, but just as importantly, political union (*l'union politique*)." A.-M. Lecoq, *François Iᵉʳ imaginaire: symbolique et politique à l'aube de la Renaissance française* (Paris, 1987).

[75] Sallust, *The Histories*, BC VI: 1–3.

[76] Ibid., *Jugurtha*, X: 6.

[77] See Lecoq, chapter 11, esp. 427–433. For the text of the entries cited here, see Guenée and Lehoux, *Les Entrées royales*, 65.

[78] Ibid., 277.

[79] In addition to Lecoq, see Théodore Godefroy, *Le ceremonial françois* (Cramoisy, 1649), 694.

[80] See Wintroub, *A Savage Mirror*, esp. chapter 3.

The Eucharistic associations with the heart should not be discounted here. Indeed, the heart, like the Eucharist, was definitive of the translation from the individual to the community, the provinces to the center (i.e. Rome), the material world to the unchanging heavens; individual acts of consumption (communion) – or the witnessing of the Eucharist's elevation – were tantamount to the individual's incorporation (paralleling the host's transubstantiation) into a larger Christian community. In this sense, perhaps the Eucharist can be seen as a kind of map, a means of making tangible the unimaginable. In the case of the map, distances beyond perception could be perceived, in the case of the Eucharist, the individual could access universal concord and *communitas*.

These ritual enactments of concordia had, of course, textual analogs. Guillaume de La Perrière, for example, employed imagery of concordance in his *Mirroir politique* (1555), depicting the estates of France – a priest, a knight, a merchant, a peasant, and an artisan (armed with a square and a compass) – surrounding a heart carrying within it an ideal city, a city governed by a "union of hearts and *l'amour commun*."[81] A generation earlier, a contemporary of Parmentier's, François Demoulins, included in his 1516 manuscript on the Psalm, *Dominus illuminatio mea*, an image of the estates of France being united by concordia (represented by Louise de Savoie), mother of *Paix pacifique et Union paisable* (that is, François I). Given the currency of this theme in sixteenth-century France, it should come as no surprise that Parmentier made concordia, the center point and refrain of his chant royal, *"L'hystoire catylinaire: Au Temple Sainct de Divine Concorde."*[82] Parmentier thus poetically describes Catiline and his followers as "mutineers" who were "full of envious hate, more dangerous than venomous serpents and lizards whose bite delivers certain death." His choice of words – *mutyn* – is perhaps significant, insofar as it once again indicates a site of translation of his life as a ship-captain and a sailor struggling to impose his authority over ship and crew into moral and political terms having to do with the government of municipalities (Dieppe) and empire (whether Roman or French). Ultimately, the conspiracy was foiled in Parmentier's account by God acting through the Virgin's mediation. As he explains in his chant's final stanza, the temple of divine Concord is the Virgin Mary, and it is she who defeats Catiline (the devil). Parmentier's ideas thus display close connections to – and familiarity with – royal propaganda and civic ceremonial, not only in his giving the notion of Concordia such primacy, but

[81] Lecoq, *François*, 430.
[82] See Chapter 3.

in articulating it as a political theology. "Of all societies and companies," he says to Ango in the dedicatory preface to his translation of *L'hystoire catilinaire*,

Nothing is so constant, and so excellent, and of such great worth as that which is made by the show and joining together of men of good faith, [with] the appearance and unanimity of good and virtuous mores. …Such company assembled in the confines of villages and cities, engenders an inviolable amity and equitable justice, so constant that in the end all will be very happy. By this, as Pythagoras wanted, many form themselves into one (*pluralité se form unite*), such as the virtue of each individual tends to amplify and increase that which we call the public interest without thinking about the excesses of particularity.[83]

As we have seen, Parmentier conceived of his translation as a means of promoting this ideal:

If in support of such virtue, I want to give and offer you [Ango] some little service, I pray you not to think only of adulation…, because inflexibly I would like to lend support to the true disciples of political virtue and, by contrast, in my fantasy to repulse all the age old villains of avaricious particularity. And this is why [I ask you] to consider my heart, and receive agreeably, in the cordiality of your benevolence, this present work; which is the *Catilinaire* of Crispus Sallust.[84]

Consider his heart, but also his mind – La Pensée – for it was in their combined action that concordia could be established and the word of God realized. For Parmentier, God's word was "engendered in the heart and thought of man,"[85] and just as the heart would hold in its universal embrace the vagaries of the terrestrial realm, so too would reason – the rule of thought (*la pensée*) – joined to and consecrated by faith (*le sacre*), reinstate concord between different social classes and different peoples, whether on Parmentier's ship, in Ango's Dieppe, in the royaume de France, or indeed, in the entire world. Thus were inner states to be

---

[83] "N'en est point une si constant et sil excellente et de si grand efficace comme est celle qui est faicte par la semblance et conjunction des homes de bonne foy, semblables et conformes en bonnes et vertueuses meurs. Et tell compagnie assemblée dedans le circuit des villes et citéz, engender une inviolable amytié et une justice pleine d'équité, si constant que la fin en est bien eureuse. Par laquelle, comme vault Pythagoras, d'une pluralité se form unite, de sorte qu'en la vertu d'icelle, chascun tend à amplifier et accroistre ce que nommons la chose publicque, sans penser à l'excés de particularité." Parmentier, *Oeuvres*, 117–118.

[84] "Si en faveur de telle vertu, je te vueil donner et offrir quelque petit service, je te prie non penser qu'adulation y ait lieu, car inflexiblement je vouldroye soubstenir les vrays suppotz des vertus politiques et, par contraire, repulse en ma fantasie tous vielz vilains plains d'avaricieuses particularitéz. Parquoy, considere mon *cueur*, et receoy agreeable, en la *cordialité* de ta benevolence, ce present opuscule: c'est le "Catilinaire" que feit Crispe Salluste." Ibid., 119, my emphasis.

[85] "La parole est engendré au Coeur et en la pensee de l'homme." Crignon, *Oeuvres*, 44. See Chapter 6.

transposed into states in general, and the microcosm translated into the macrocosm by a host of many men acting in perfect synch and disciplined proportion – to scale not only the world's disparate geographies, but theological and political order.

The distinction between rhetorical performance and efficacy, must have been quite clear on the decks of La Pensée when Parmentier read his *Oration* to his unhappy, unruly, and potentially mutinous crew. Hence, the medico-theological ritual of dissection which bloodily underlined Parmentier's authority with the threat of force. Similarly, despite good intentions – and the stated desire for concordia amongst the different estates of France – discord, whether religious, political, or social, was met in no uncertain terms with extreme violence. Criminals were tortured, dismembered, and put to death; the poor were threatened with similar punishments, as were peasants who resisted taxation or men and women of any class who challenged religious orthodoxy. Good intentions and the desire for concordia were always, it seems, backed up with force. As with Parmentier's publicly performed dissections, torture had religious dimensions. For Parmentier, the violence done to his men's bodies was both an investigation into the causes of disease and an exploration of God's creation through the microcosm of man. Torture of the dead forced the body to reveal its secrets while simultaneously serving to make the living (and suffering) bodies of his crew docile. For political and ecclesiastical hierarchies, torture was a means of cleansing the body politique, and making the heart a fit receptacle for God's Word; it was also, of course, a demonstrative rhetoric of power and control. Truth, it seems, like concord, was tied closely to coercion. Perhaps this is why knowledge and expertise were so distrusted,[86] and why men like Parmentier and Crignon went to such great lengths – as did the writers of royal propaganda and the designers of court fêtes – to translate their power into mystical notions of concordia and imperial Christianity. In the chapter that follows, we will further explore the mechanisms by which concord – whether political, social, theological or epistemic – was conceptualized, maintained, and disciplined across the borders of nature and culture, self and other.

---

[86] Even today, as Richard Hofstadter famously points out, the mind is denigrated as a site of willful duplicity and deception, when compared to the pure and spontaneous truths expressed by the heart; see his *Anti-intellectualism in American Life* (New York, 1962).

5   Confidence: A Balance of Trust

Confidence is hard to come by and difficult to pin down. Somewhere between faith and credulity, it is more than just a little bit slippery. In late medieval and early modern France, trust (*confiance*) was an artifact of one's location in a geography of social, cultural, and familial relations.[1] Even total strangers, if enmeshed in these social-geographies, could be considered creditable. Kinship and patronage networks were only one aspect of this trust; belonging to a parish, a guild, going to a particular tavern or inn, participating in municipal government, playing a part in a local confraternity, all could serve to extend circles of trust and credit.[2] The constraints – the shared norms and expectations – offered by these community ties were important resources in managing epistemic credibility, social standing, and on-going (tacit or explicit) contractual relations. Trust in this sense was an artifact of membership in – or association with – kinship, social, or institutional networks.[3] As trade widened and circles of family, friends, and associates attenuated, the work of securing the currency of trust relied on written representations, testimonies, tokens, oaths, seals, signatures, and hearsay.[4] They also relied upon acts of self-representation that could attest to one's trustworthiness; these confidence-building

---

[1] See, for example, Frisch, *The Invention*, and Shapin, *The Social History*.

[2] Sheilah Ogilvie Cesifo, "The Use And Abuse Of Trust: Social Capital and its Deployment by Early Modern Guilds" Working Paper No. 1302 Category 10: Empirical and Theoretical Methods (October 2004); Beata Kumin, "Useful to Have, but Difficult to Govern. Inns and Taverns in Early Modern Bern and Vaud," *Journal of Early Modern History* 3 (1999), 163; and James Murray, "Of Nodes and Networks: Bruges and the Infrastructure of Trade in the Fourteenth-Century," in Peter Stabel, Bruno Blondé and Anke Greve (eds.), *International Trade in the Low Countries, 14th–16th Centuries* (Leuven 2000), 4–5.

[3] See Francesca Trivellato, *The Familiarity of Strangers: The Sephardic Diaspora, Livorno, and Cross-Cultural Trade in the Early Modern Period* (New Haven and London, 2009).

[4] On the epistolary extension of trust see, for example, Suze Zijlstra, "To Build and Sustain Trust: Long-Distance Correspondence of Dutch Seventeenth-Century Merchants," *Dutch Crossing: Journal of Low Countries Studies* 36 (2012): 114–131; Franz Mauelshagen, "Networks of Trust: Scholarly Correspondence and Scientific Exchange in Early Modern Europe," *The Medieval History Journal* 6 (2003): 1–32.

credentials could include signs of distinction such as appropriate dress, manners, and speech.[5] However, these signs, insofar as they were learned dispositions, affected behaviors and performed persona, were – or could be – construed as untrustworthy performances, hence the proliferation of sumptuary laws, the general distrust of rhetorical schemes and tropes, anti-courtier satire, and the suspicion directed toward adopted mannerisms (or indeed, opposing prescriptions advocating *sprezzatura* and plain speaking), all of which point to the fragility of the frontiers separating the trustworthy from the scoundrel, the genuine from false.

Trust given could easily be betrayed. Even a casual examination of archival records from this period indicate an explosion of litigation over issues arising from broken trust and misplaced confidence.[6] This seems to point to a devolution of trust away from qualities possessed – or attributed to – individuals, to confidence in the impartiality of adjudicating (constraining) institutions – e.g. the rise of courts administering, laws governing, and bailiffs enforcing the terms and conditions of trade. In early modern France, however, the law and its administration were tangled affairs of overlapping and competing local, regional, and national jurisdictions; it was often corrupt; and it was far from bureaucratically neutral and rationally efficient. Indeed, the conclusion that the law was an expression of the interests of the powerful should come as no surprise, nor should the fact that there were many different kinds of power and therefore many different and competing laws and jurisdictions. As Montaigne put it, French laws are "so confused and inconsistent" that

---

[5] "The beginning of the modern era, Elias argues, was a moment of change and uncertainty between two ages of social glaciation. The unity of Catholicism had broken down, and the rigid hierarchies of the Middle Ages had suffered profound damage as courtly and chivalric society was called into question, but absolutism had not yet established its dominion. This was a period of social and cultural realignment. Social groups were more diverse than ever before, and relations among them more complex. Changing societies required a new common language and common points of reference." See Jacques Revel, "The Uses of Civility" (167–205) in Philippe Ariès and Georges Duby (eds.), *A History of Private Life*, volume III, Roger Chartier (ed.), *Passions of the Renaissance* (Cambridge, MA, 1989), 173. With regards to the relationship between social credibility and epistemic trust, see Shapin, *A Social History*, and with regard to its specifically rhetorical aspects see Shapin and Schaffer, *Leviathan*, and Wintroub, "The Looking Glass of Facts: Collecting, Rhetoric and Citing the Self in the Experimental Natural Philosophy of Robert Boyle," *History of Science* 35 (June 1997): 189–217 and, especially, Juan Pimentel, *Testigos del Mundo: Ciencia, literatura y viajes en la ilustración* (Madrid, 2003).

[6] As Elise M. Dermineur notes, "there are extraordinary and extensive collections of judicial court records, accounting for thousands of kilometers of shelving in archives all over France. These are undeniably the largest resource for early modernists. Civil lawsuits account for most of the judicial proceedings while criminal records represent only a ratio of 1:20." Elise M. Dermineur "The Civil Judicial System in France," *Frühneuzeit-Info* 23 (2012): 45–52.

they invite "disobedience and faulty interpretation, administration and observance."[7]

Other, extra-legal, kinds of precautions were also employed to adjudicate credibility and maintain and regularize trust: the standardization of weights and measures, for example, promised to materialize the foundations of credit. Estienne, citing Paulus, defined *creditum* as *"faire prest et credit, et suyvir la foy d'autruy. Et se dict des choses qui se prestent consistans en poix, nombre, ou mesure: lesquelles ne se rendent en mesme espece, ains suffist en rendre de pareille en nombre, poix ou mesure."*[8] Knowledge of numbers – of weights and measures – could, it was thought, protect one from deception by building material constraints to channel and maintain trust.[9] Thus, on the one hand trust was a sentiment or a value; on the other, it was an observable quality – something quantifiable that could be held in the hands and measured. But the idea of an impersonal commercial world based on standards of measure and numeracy – that is, on a kind of objectivity of the market located in commonly held standards of weight, measure, and accountancy – was far from reality in the sixteenth century.[10] In early modern France, measurement was much like the law; different weights and measures proliferated, not only from one country to the next, but between different cities in the same country and between villages only miles apart.[11] Even money differed from town to town and city to city. In Normandy, for example, in addition to French money local currencies included English, Spanish, Portuguese, and Flemish coin. For this reason contracts from this time frequently specified not only the terms of a transaction, but also agreed upon standards of weight and measure, and the kind of currency that was to be used.[12] The proliferation of foreign coin continued despite repeated attempts to restrict it. This metrological anarchy was a reflection of the weakness of central authority's ability to impose uniform standards – whether in the law, in weights and measures, or in money; on the other hand, it also attests to the strengths of countervailing local interests that resisted – and/or appropriated and inflected – the extension of royal (or municipal) power.

---

[7] Montaigne, *Essays*, III:13, "Of experience," in Frame's edition, at 821.

[8] See Estienne, *Dictionarium*, s.v. *creditum*.

[9] See, for example, Claude Platin's translation of Juan Ortega's, *Oeuvre tressubtile et profitable de lart et science de aristmeticque: et geometrie translate nouvellement despaignol en francoys* (Lyons, 1515), A iii^r, as cited by Natalie Zemon Davis in the "Business of Arithmetic."

[10] Colbert's regulation of markets over a century later aimed at creating this sort of neutral ground of confidence wherein trade amongst strangers could take place; it thus played to the interests of merchants while at the same time furthering the power of royal – state – bureaucracy. See Philippe Minard's *La fortune du colbertisme* (Paris, 1998).

[11] See, for example, Ronald Edward Zupk, *Revolution in Measurement: Western European Weights and Measures Since the Age of Science*, Memoires of the American Philosophical Society, Vol. 186 (Philadephia, 1990), 3–24; more generally, see Wintroub, "Translations."

[12] See Mollat, *Le commerce*, 385–387.

Even when common standards could be agreed upon, instruments remained untrustworthy, as did the eyes and hands of those who used them. Merchants, for example, were frequently regarded with mistrust, and characterized as "*menteurs, deceptifz, fins et subtiles, à tromper inventisz.*"[13] Their weights were said to vary with their interests, sometimes lighter, sometimes heavier. The Portuguese famously viewed the weights and measures of the Africans they traded with on the Gold Coast as satanically inspired fetishes, and as a kind of witchcraft. Back at home, mathematical practitioners (like Oronce Fine and John Dee), and their instruments, were often viewed the same way.[14] Like the "savages" with whom the Portuguese traded thousands of miles away, practical mathematicians risked being taken for magi dabbling in the black arts, or as simply rude and socially inferior mechanics. Similarly, their instruments were thought to be dangerously uncanny innovations that held frightening occult properties – how else could a compass always point north, or an astrolabe guide a ship across unknown seas? Thus material registers, along with their numerical translation, were not necessarily conducive to trust; sometimes, and in certain situations, they could have precisely the opposite effect.

Despite the reliability (or not) of empirical measures of trust (as for example, those possessed by Parmentier and Crignon in charting their ship's course to Sumatra), or the trustworthiness and honesty of those who utilized them, other factors could always intrude to betray a trust given, whether recalcitrant natives, inclement weather, worm-eaten wood, hostile microbes, pirates, or a mutinous crew. The means to manage, or at least mitigate, these potential risks relied on further acts of quantification made with more metaphorical tools – for example, instruments of "credit" rather than those made of iron and brass: namely, joint-stock associations where the price of a voyage could be underwritten by multiple partners,[15] and by maritime insurance and letters of credit backed up by courts of arbitration.[16] These aptly named credit instruments,

---

[13] Arden, *Fools' Plays*, 117.

[14] See Adam Max Cohen "Tudor Technology in Transition," in Kent Cartwright (ed.), *A Companion to Tudor Literature* (Sussex, 2010), 95–110.

[15] For example, the company that was formed to sponsor Verrazano's 1524 expedition to a "lieu nommé les Indes en Kathaye"; see Michel Mollat and Jacques Habert, *Giovanni et Girolamo Verrazano, navigateurs de François Ier* (Paris, 1982), 118; see also Chapter 1.

[16] These innovative instruments of credit were much more developed in Flemish cities and among Mediterranean merchants, but were increasingly utilized by the Spanish and the Portuguese. The French did not begin to catch up until the mid sixteenth century, primarily through the auspices of Spanish merchants residing in Rouen and through the mediation of Italian banking families based in France. In any case, in the period under discussion, these instruments were at best primitive and did little to mitigate the dangers of long distance navigation.

however, were much like their more material counterparts: a gamble of trust and a leap of faith, not only that a ship would return with the merchandise promised, but that one's partners and/or underwriters were – and would remain – both honest and solvent.[17]

Trust given on the basis of these new forms of hands-on mathematical instruments, like trust in instruments of credit, was not *sui generis*; it derived, like most things, from good packaging and an audience that wanted to be persuaded. Parmentier's appeal – his credibility – for potential partners in his voyages, for example, was that of a spokesperson for a loose-knit confraternity of provincial poets, humanists, and merchants. His poetry made clear the connection between God, voyages of exploration, commerce, and the skills of a poet. It also addressed and actively engaged a like-minded audience in a bid to speak for – and be exemplary of – their cultural and spiritual aspirations and interests. Trust in Parmentier and his methods, like acclaim for his devotional poetry, was trust in his eventual success and thus credit with the very people who were most invested (in every sense of the word) in overseas exploration and trade.[18] Parmentier's crew's lack of faith in his abilities, and his need to reassure (or coerce) them is indicative of the precariousness of trust based on material measures alone. It perhaps goes without saying that these measures were entangled in – produced by and productive of – articulations and assertions of status hierarchy. Parmentier's *Oration* and his postmortem anatomies must surely be seen in this sense, that is, as attempts to shore up both his expertise and his authority in the eyes of his men, who had as little reason to trust in the new-fangled instrumental means of mathematical navigation, as they did to trust in the justice of their social "superiors" who had persuaded them to sail on rickety ships across unknown seas. Navigating men, it seems, was every bit as difficult as navigating the high seas.

So what was this word "confidence" that seems to have no real referents other than faith or credulity. Robert Estienne, for example, made confiance both performative, citing Terence: "we believed him;" and provisional, citing Plautus, "we will lend it to you." One holds trust – like a token or coin – but then lends or gives it away: it thus migrates from the giver to become the possession of the receiver, namely, as credibility. Others can inspire trust and confidence; but trust is something that is

---

[17] Misplaced trust found recourse in merchant associations (and later) merchant courts and special maritime courts, for example, the *amirautés* of Dieppe and Rouen. Though Dieppe's maritime court existed from at least the early sixteenth century, the records are very incomplete; virtually all sixteenth-century archives having been destroyed by English bombardment late in the next century.

[18] See Wintroub, *A Savage Mirror*, esp. chapter 4.

ours to give. We trust somebody because we credit them, that is, we lend them credibility as if we were lending them money. However, once credit has been given, we still, in a way, possess it, or at least a trace of it, like a promissory note guaranteeing a return – an ultimate act of redemption.[19] Thus trust was (and is) a kind of seesaw balance of possession – a circulation between self and other in a social tautology of mirrored transactions of credit and debt. Indeed, how could this game of confidence ever be more than simple credulity: a naïve mistake, an error in judgment, a miscalculation, an illusion? In other words, how could it be faith?[20]

Trust could often be won by appealing to, or articulating, common values and ideals. Institutions, such as the Church, for example, had advantages in this regard, as they embodied a certain choreographed inertia of ritual practices and dispositions. In other contexts, trust could perhaps be said to be the result of competition or negotiation: something that could be won or lost, given or taken away at any moment by either side of a transaction. Such battles, of course, also occurred on the institutional level, though the stakes were in the first instance local, and then, incremental. More generally, in transactions requiring trust, some clearly had more credit than others. This variability can, in theory, be mapped onto social and ontological hierarchies: savages, lower-class criminals, strangers, women, Jews, and heretics had very little; learned elites, important merchants, nobles, and the clergy had somewhat more.[21] Though, this too was variable, and a question of circumstance and perspective. It was not only the poor, the heterodox, the Jew, the lowly sailor, and savage who had little cause to trust in the word or expertise of their social "superiors"; the self-confidence of the *nouveau riche* trying to become noble, or the noble trying to maintain status in the face of derogation, or indeed, either, as well as the clergy, in the face of escalating confessional conflicts, were also challenged and plagued with doubt.

Winners in contests of trust, of course, did not need to defeat their opponents outright; rather, they could try to co-opt and enlist them in the furtherance of their own arguments and interests. Parmentier's and Crignon's theological verse comes to mind. Nevertheless, credibility was – as the shipboard postmortems attest – difficult to acquire and maintain. Alternatively, giving away one's trust too easily was a risk that could open one up to all sorts of trouble. Allies could easily turn coats, and become powerful enemies. "Confidence," in this sense, is not

---

[19] Mauss, *The Gift*, esp. 8–11.
[20] I have discussed this in "Taking a Bow," for example, at 787.
[21] Insofar as they were entangled in networks that articulated – with varying degrees of specificity – social and institutional hierarchies that stored and extended credit.

only something felt or adjudicated in response to others, but refers to an opening up, a revelation of one's own heart in a token, a gift, a confidence that demanded a counter recognition, a balance struck, in the gift of trust. As the *Dictionnaire de L'Academie* put it: *"Parler à quelqu'un en confiance, pour dire, Luy parler à coeur ouvert; & comme estant bien assuré qu'il gardera le secret."*[22] However, a confidence could always be betrayed, and for any number of reasons: for joy in the act of treason itself, as "innocent" gossip, a slip of the tongue, or because one wanted to change sides and prove one's credibility – and worth – to another. The negative qualities associated with confiance, as in the confidence men of the nineteenth century, had deep roots in the word. Mistrust, betrayal, and treason were always lurking just behind. Cotgrave's dictionary, for example, includes the common variant of confiance, the *confidentiaire*, not only as someone entrusted with a confidence (important, private and confidential knowledge), but someone who is "treacherous, faith-breaking, trust-deceiving."[23] To believe, to have faith in, or to have confidence, thus carried within it the profoundly troubling qualities of its opposite – naiveté, foolishness, and, according to Cotgrave, "docility."

How then does one credit someone or something as reliable and trustworthy? By what measure could honesty be adjudicated and dishonesty punished? How could one confidently approach strangers who could not be vouchsafed by any accepted criteria of reliability and trustworthiness? What was the measure of trust and how could it be maintained? The voyage of La Pensée and Le Sacre will once again guide us as we pursue the practices, skills, and improvisations that constituted the precarious balance between trust and betrayal, profit and loss, life and death. By following the trajectory of Parmentier's ships as they crossed perilous waters to meet and trade with unknown peoples, we will encounter – across the frontiers of social hierarchy, Man and Nature, "Europeans" and "Others" – strategies for negotiating trust, whether between officers and crew, ships and seas, or French merchants and Sumatrans.

A Kid and a Posse

On April 25, 1529 Pierre Crignon took the height of the sun. It was 16 degrees and 8 minutes. They dropped anchor off the coast of an unnamed island in the Cape Verdes. The next day, four boats, two from each ship, rowed ashore, each manned by 20 "well armed men in good order" to

---

[22] *Dictionnarie de l'academie.* s.v. *confiance.*
[23] Cotgrave, *A Dictionarie*, s.v. *confidentiaire.*

look for water and supplies. Their arrival surprised a group of Moors and their Spanish overseer. Trust in such a situation was a decision of life or death, and it was better to be safe than sorry; the Spaniard and the Moors made for the hills. Before they were out of earshot however, the Portuguese from the Sacre shouted out that they were the "army of the French navy equipped for war on its way to the Antilles"; they had lost their fleet (*notre bande*), he said, and they were just looking for water and other refreshments (*refreschissemens*), if there were any to be had.

The lie is interesting. It was performed for a reason and is revelatory of motives, perceived needs and intentions.[24] Put another way, when strangers meet, the lie performs an introduction. Honesty, opening one's heart, could be extremely dangerous. When trust is lacking, anything can happen. In this case, the lie that was told was meant to intimidate – to stake out the high ground in a relationship of potential conflict, where bluster and a show of strength might persuade – indeed compel – others to accede to one's demands. In other words, it was a tactical threat; but it was also a test: was this "Spaniard" more interested in his countrymen's possessions across the Atlantic in the Antilles (with whom the French were supposedly going to war), or in his own skin; if the latter was the case, then surely he would want to get them underway and out of his hair as quickly as possible. The lie, of course, also had another purpose: namely, the French were not about to reveal their true intentions – to search for spice and gold in places and seas claimed by the Portuguese – to a stranger. Though the European with the Moors is identified as Spanish, Crignon had no way of knowing this for sure; or indeed, if he could be trusted at all.

The "Spaniard" replied to their request with something of a non sequitur, stating that some 12 leagues away there were two Portuguese ships that had been pillaged by Bretons on their way from Madeira. Was he calling the French out? Giving them the lie and standing his ground, at least rhetorically, before the superior force presented by these French "pirates"? And were they the Breton pirates who pillaged the two Portuguese ships? If so, what were their real intentions? Was it really water they wanted or was it plunder?

A few days later, Crignon observed the smoldering fires coming from a volcano on a neighboring island. The only volcano active in the region was on Fogo (the island Crignon identifies as Fuegos). We can thus place this meeting definitively on the island of Santiago (whose

[24] Timothy R. Levine, Rachel K. Kim, Lauren M. Hamel, "People Lie for a Reason: Three Experiments Documenting the Principle of Veracity," *Communication Research Reports* 27 (2010).

latitude is approximately 15 degrees and 4 minutes, about 80 miles away from the position calculated by Crignon by modern reckoning). Surely the port alluded to by the Spaniard was Ribeira Grande (now Cidade Velha); the first European settlement in the tropics, a major node in the Portuguese slave trade, and an important port of call in the *Carreira da Índia*.[25] Behind the Spaniard's retort was not simply someone who was outnumbered and outgunned, but someone who had very up-to-date intelligence about whose ships were doing what in a Portuguese (an enemy) port several hours sailing away. There was surely more to this Spaniard than met the eyes. Was he a spy or simply a well-informed foreigner living thousands of miles from home on a small island off the coast of Africa in the midst of Moors and in close proximity to an important Portuguese port?

We can well understand why Parmentier and his men did not want to reveal their true objectives to just anyone. Indeed, what would the Portuguese do if they learned of their true intentions? Conversely, the Spaniard had even less reason to trust the French. There was no love lost between the Spanish, the French and the Portuguese, especially in their colonial outposts on the coasts of the Atlantic and Indian oceans. In the 1520s, French North Atlantic pirate-merchants terrorized shipping – pillaging vessels from the coasts of Portugal to the New World.[26] The reputation of the French as rapacious pirates had surely preceded their arrival at this remote outpost off the African coast.

The Spaniard led them over difficult terrain to water. While the barrels were being filled he told them that he intended to commit treason (*alloit faire quelque trahison*) and search out men to stop them if they didn't quickly return to their ships. He must have been a very nervous man – the bluster of his threat indicates his complete lack of trust in the French. We need to put this threat into context, however. It began with a preamble – an offer of a possible gift of food – the promise of a kid (*un*

---

[25] See B. W. Diffie and G. D. Winius, *Foundations of the Portuguese Empire, 1415–1580* (Minneapolis, 1971), vol. 1: 377. Though the treaty of Zaragoza would be signed only a few days after this encounter, on April 29, setting the eastern meridian dividing the world in favor of the Portuguese claim to the Mollucas, the Spaniard could not have known this; and indeed, despite the assurance made by the treaty, the Spanish continued to compete with the Portuguese in the region for years to come.

[26] Piratical acts conducted by Ango's men were, at times, given the legitimacy of royal policy through the granting of letters of marque by the French King. The Spanish and Portuguese responded in kind. In most instances, the fight for freedom of the seas carried on by Normandy's merchant-pirates was subordinated to François' transalpine ambitions and his enmity for Emperor Charles V. This message was clearly ignored by the likes of Ango who pursued his own interests regardless of the sanction of the crown (though if he could, he certainly would – and often did – co-opt the King's desires for his own purposes).

cabry) to be sacrificed to the crews' hungry bellies in exchange for their leaving. The conditionality of the promise is perhaps indicative of the Spaniard's place in this interaction and to his intense apprehension: he was, he said, going to look into the possibility of finding a cabry; that is, he said he would make enquiries, but he wasn't entirely sure he could deliver. It is at this point, that he shifts from making a promise, either a gift of food or of sincere intentions to do so, to making threats. Could he carry through with this "treason"? Perhaps. Who could tell what or whom he knew, or what kind of damage his for now hidden allies could do if brought into the fray?

But the wind was blowing and the seas were rough and they were in peril of losing their boats; the French couldn't leave. While they anxiously waited, they sighted the Spaniard descending from the mountains with the promised kid. The signal man, the *port-enseigne*, gestured for him to approach, but he didn't "dare"; the port-enseigne waved a shirt that he wanted to give in trade, but he still wouldn't come. So accompanied by others (*quelques autres*) the signal man went to the Spaniard, giving him two shirts in exchange for the cabry. The Spaniard gave them the kid saying he could meet them the following day on the northern side of the bay where he lived; there, he said, he had cows, chickens, and water. It seems that his excursion in difficult terrain that he knew (and the Frenchmen did not), along with the gift of food, was a test and a gesture – a means of better getting to know the French and their intentions. This was ground he knew and into which he could easily slip away, and maybe the shirts also seemed of good quality; it seems the French had passed muster; perhaps they weren't out to get him after all. He thanked them *grandement* and left. But if it were the case that he trusted the French after helping them fill their boats with water, why did he then threaten them with a possible treason, or offer them more gifts? And why was he so wary that he wouldn't approach when the signal man gestured? His decision to invite them to meet the next day seems perplexing given the nature of their interactions thus far.

The men from the Sacre and the Pensée were stuck on shore for many hours; they struggled against the wind and uncooperative seas and were only able to rejoin their comrades late that night. The next day, April 27, they returned to land and were greeted by the Spaniard and 10 or 12 Moors armed with pikes and crossbows. They welcomed the French (*firent bon accueil*), who were directed to water, beef and chickens. When the Normans insistently tried to pay their host (two *écus* was the price they offered), he would hear nothing of it. He said, rather, that he prayed to God to grace them with a good voyage, telling them that if they were ever to return they would have many fine gifts, and he was "vexed" that he couldn't do better for them now, on such short notice.

Crignon now takes the time to tell us a little more about his host, recounting how this Spaniard was the leader (*maistre*) of all the others because "he commands, and the others obey." He had three or four native wives or daughters who served him. After this brief introduction, he lets the Spaniard speak for himself about how his wife had wept the previous night, "thinking that our men were going to take or kill him."

What was the turning point in this interaction; when did suspicion become confidence? When and how did trust happen? Clearly, for the Spaniard, trust was secured by arranging for an armed escort to be deployed as a counter to any potential threat. In other words, he invited the French to visit his home knowing, despite his wife's tears, that he could assemble allies sufficient to ensure his safety. Here is a credible answer to our earlier question about the constitution of trust. Trust required a kind of balance of power.[27] This was true, as we will see, both in respect to relations with others, as with relations with nature.

Anchors Aweigh

On April 27, at about 5 pm, they hauled anchor and put to sail and followed the wind to a cape on the south of the island for the night. In the night's second watch they saw great eruptions of fire from atop the island to their west. Crignon compared the flames he saw to Sicily's Mount Etna. It would be 71 days before they were to set foot on land again.

Life on board ships like Le Sacre and La Pensée was hard. The decks were crowded and the sun was unrelenting. They were far from home, sailing perilous seas, with no idea when, or indeed, *if* they would ever set foot on land again. And if they did, who could predict what kind of reception they would receive? Only three years before, another expedition to the Indes sponsored by Ango was attacked upon reaching its destination in Sumatra (in Aceh). The pilot was killed alongside much of the crew. The survivors, including their captain, Pierre Caunay, fled back to France only to be shipwrecked off the coast of Madagascar. They managed to make their way to the African mainland, but were taken prisoner by the Portuguese. They were never heard from again. But there were also more immediate problems than what the future

---

[27] It is unclear, however, despite trust won, how secure or sincere it was; it is thus very doubtful that the French would ever have disclosed to their new Spanish "friend" their true objectives. Coincidently, the expression, "balance of forces" was introduced into political discourse by Bernardo Rucellai (discussed in Chapter 1). On Rucellai, a relation of Ango's co-financiers, see Felix Gilbert, "Bernardo Rucellai," 102.

might bring: hunger, thirst, fatigue, and disease were constants. The work was backbreaking and the conditions were dangerous. The tangible uncertainty of their voyage had to weigh heavily on the crews' hearts and minds. While Parmentier appealed to his poetry to reassure his men, Crignon employed other means.

The voyage of La Pensée and Le Sacre was charted in his log by days and months; it was also harnessed to the trajectory of religious fêtes – the days of saints and holy days: Easter, the eve of Saint Nicolas, Pentecost, the Trinity, the days of Saint Pierre and Saint Paul – linking their ships' physical voyage across space to a spiritual order that would inspire confidence and faith. However, a more regular, constant and tangible mapping onto celestial coordinates was also enacted by the physical presence of Crignon, La Pensée's astrolabe-armed cosmographer, taking the height of the sun each day at noon. A great deal of performative work went into producing the simple declarative phrase found so often in his Log: "the height of the sun was taken at noon." Indeed, Crignon's daily readings were conducted not only as a means of locating La Pensée in an abstract coordinate system derived from the stars, they were also a ritual meant to inspire and reassure her crew that they were being led to their intended destination by someone who knew how to get there. Not a man, but a conduit for divine knowledge, Crignon and his instruments, maps and declination tables, along with his expertise, were a performance of his authority and his unique occult power to see (instrumentally and mathematically) where others could not to chart the course of La Pensée simultaneously in both the heavens and on the seas. As he said in one of the songs he presented at the Puy de palinod,

> Car le patron sçait si bien gouverner
> Par l'astrolabe et compass serieux
> Qu'ilz ont trouvé sans long temps sojourner
> L'isle où la terre est plus haulte que les cieulx.[28]

This grand patron, the "pilote tressage" who "composa la grand cosmographie" in this poem was Almighty God, and Crignon on board La Pensée was surely his instrument. But the translation from the heavens to ships powered by the winds, and from paradise to their destination in the Indes, was by no means unproblematic. As with any ritual performance, a great deal of work had to be accomplished behind the scenes before it could take place. A balance had to be struck between the forces of nature and the assemblage (l'escale) of men, instruments, and technologies on Le Sacre and La Pensée.[29]

[28] See note 76, Chapter 2.
[29] See especially John Law, "On the Methods."

The height of the sun needed to be taken at its maximum altitude, at noon. To determine if it was noon, Crignon might simply have estimated when the sun was at its highest point, or he could have tracked it with repeated measures with his astrolabe, taking the reading where the sun was at its highest point, or he might have used some form of gnomon compass, such as a diptych dial or astronomical compendium, which would have been easier to use than an astrolabe for this purpose. This in turn would have required the maintenance of the compass needle, making sure that it was regularly touched by a lodestone (a *pierre d'aimant*) to sustain its occult properties. Additionally, he would have needed to calibrate the compass to compensate for magnetic variation from true north. This was important not only for orienting gnomons and finding meridians, but for setting the ship's bearing. This could have been done by taking the sun's shadow slightly before and slightly after noon to determine its azimuth, comparing this with the azimuth found with the compass' magnetic needle, and then adjusting the compass rose relative to the needle to reflect true rather than magnetic north.[30] Given the importance that Crignon attached to being an expert cosmographer and navigator, this is perhaps something he did, especially since he wrote a now lost manuscript on the variation of the compass needle. Indeed, one can well imagine him making quite a show of his careful observations, and of fiddling about with dials and books at or around the noon hour.

It is noteworthy that Crignon's journal also frequently indicates position by longitude. This was a much trickier proposition. Longitude could not be calculated on board a ship with any degree of accuracy until the end of the eighteenth century.[31] Were the longitude measurements in Crignon's Log just part of the show – an attempt to provide an illusion of certitude while sailing on unknown seas with men who were tired, hungry, and scared? This is unlikely; while instruments, books and fancy talk might have impressed, the details of an abstract Ptolemaic coordinate systems of latitude and longitude noted in his log would surely have gone over the heads of most of the crew, even if they could read. Moreover,

[30] See E. G. R. Taylor, "Jean Rotz: His Neglected Treatise on Nautical Science," *The Geographical Journal* 73:5 (1929): 455–459. Horace Perry, *The Age of Reconnaissance* (Los Angeles and Berkeley, 1981), 97. As suggested by Pedro Nunes (1502–1578), a *converso* physician and professor of mathematics, known not only as the greatest cosmographer-mathematician of his day (responsible for important and widely influential works on cartography and navigation), but as a translator, commentator of and lecturer on Greek rhetoric.

[31] See, for example, Derek Howse, *Greenwich Time and the Discovery of Longitude* (Oxford, 1980); on the difficulties of determining longitude at sea for a much later period see, for example, J. A. Bennett, "The Travels and Trials of Mr Harrison's Timekeeper" in Bourguet, Licoppe and Sibum (eds.), *Instruments, Travel and Science: Itineraries of Precision from the Seventeenth to the Twentieth Century* (London, 2000), 75–95.

pilots and cosmographers, even self-assured ones like Crignon, had to know that their estimates of longitude were deeply flawed.[32] This said, Crignon surely had faith that his calculations would aide in guiding them to their destination. Indeed, however inaccurate, he must have believed that his position estimates were better than simply guessing.

In theory, longitude could be found by lunar observations relative to that of a star or the sun (the lunar distance method), which would then be compared to ephemerides that set out the relative positions of these celestial bodies at local time at or near their point of departure. Vespucci, for example, claimed to have done this in 1499.[33] Closer to home, Jean Rotz, Crignon's contemporary from Dieppe, claimed to have used his cross-staff at dawn to measure the distance between the sun and the moon, which he then compared to ephemerides for Ulm to determine his longitude, which placed him somewhere, or so the numbers tell us, in the vicinity of the Moluccas.[34] Wherever he was, it is highly doubtful that either he, or Vespucci, calculated longitude at sea in this way. More likely than not, the difficult-to-make instrumental observations and lengthy mathematical calculations included by both Vespucci and Rotz were simply textual performances of the latest navigational theory composed to impress their powerful patrons: respectively, Lorenzo di Pierfrancesco de' Medici and Henry VIII. Crignon, however, did not reveal his methods,

---

[32] Lopo Homem, the Portuguese cartographer, for example, wrote in an undated sixteenth manuscript that Nunes "had drawn up a master-chart of the route to India, on which longitudes had been determined by observation of lunar and solar eclipses… and that Portuguese pilots (between 1529 and 1557) had been ordered by the crown to use charts copied from it. Use of the new chart, writes Homem, had caused shipwrecks and accidents; Portuguese pilots were having their charts secretly made in Spain and elsewhere according to traditional methods and Homem therefore entreats the King to rescind the obligation to follow Nunes' chart." See W. G. L. Randles, *Portuguese and Spanish Attempts to Measure Longitude in the 16th Century* (Coimbra, 1984), 13 and 17–21.

[33] Felipe Fernández-Armesto, *Amerigo: The Man Who Gave His Name to America* (New York, 2007), 77–82.

[34] Interestingly, almost a year before Le Sacre and La Pensée were there, on January 15, 1529. See, for example, E. G. R. Taylor, "Jean Rotz: His Neglected Treatise on Nautical Science" in *The Geographical Journal* 73: 5 (1929): 455–459, and Helen Wallis (ed.), *The Maps and Text of the Boke of Idrography presented by Jean Rotz to Henry VIII now in The British Library* (Oxford, 1981), 6. Ango sent out several expeditions to the Indes prior to Parmentier's; in addition to the failed expedition of Pierre Caunay, we could add that of Jean de Breuilly, sailing on La Marie-de-Bon-Secours, who was sent to look for him, but was captured by the Portuguese in Zanzibar in 1528. Perhaps Jean Rotz was on a clandestine mission to find his lost compatriots. Or perhaps he was on board either Le Sacre or La Pensée. Some have speculated that the date Rotz supplies, January 1529, was old style and should be corrected to 1530, which would then place him in the Moluccas on their return voyage to Dieppe. This is a possibility and would help explain imagery on the maps he made for his *Boke of Idrography* which seem to depict events from Parmentier's voyage (see Chapter 6). Might he have been Jean le Peintre? I remain skeptical.

perhaps because he was not composing an account to impress elite readers. Or perhaps because he did not want to reveal his secrets until he had a chance to publish them. *La perle de cosmographie*, written in 1534 and dedicated to Admiral Chabot, is said to have argued that there was a constant scaled disposition of magnetic variation according to longitude beginning with a zero point meridian located around the Azores.[35] Following from this, his measurements of the height of the sun were perhaps used not only to find latitude, or as corrective to the compass needle's variation, but also as a putative fix of the meridian vis à vis the declination of magnetic azimuth. Nevertheless, it is far more likely that Crignon eschewed such esoteric astro-mathematical techniques to rely instead on dead reckoning to estimate La Pensée's longitude. Pursuing this theory, the latitude determined by Crignon every day at noon, along with the ship's heading (its compass bearing), and its speed over a specific span of time, taken at various interludes during the day, would have been recorded on a device such as a traverse board. These "fixes" would serve to transform individual readings, samples as it were, into patterns that could be interpreted as the ship's movement across time and space that could then be plotted on a chart and used to give an estimate of longitude. Perhaps more important than the accuracy of this method, however, was its performative effect of extending trust from Crignon and his arcane and mysterious instruments, books and maps, to the bodies and know-how of La Pensée's crew.

To determine the speed of a ship, for example, a piece of wood, a log, would be attached to a knotted line and thrown off the stern; one member of the crew would control a spool to make sure that the rope unwound smoothly, while another would count out loud as knots tied into the rope passed through his hands against a specified amount of time as measured by the length of a prayer or, more likely, an hourglass (usually a half minute glass) operated by still another. This procedure, of course, presupposed prior acts of mensuration, perhaps supervised by Crignon, or the ship's master, that marked out a consistent unit of distance, typically a fraction of 42 feet for a 30 second glass, measured out in a division of knots placed along the line. This standard, scaled to 1 nautical mile per hour, would be hammered into or marked onto the deck where it would be used to calibrate the distance between knots as ropes aged – stretched-out or shrank – with use. This in turn required prior agreement as to the length of a nautical mile and its division into

---

[35] The manuscript in which he made this claim, *La perle de cosmographie*, has been lost. See E. Leroux, ed., *Le discours de la navigation de Jean et Raoul Parmentier de Dieppe: Voyage à Sumatra en 1529. Description de l'isle de Sainct-Dominigo* (Paris, 1883), xxi–xxii n. 2.

feet, which in turn required an agreed-upon estimation of the earth's circumference, an understanding of its division into 360 degrees, a numerical measurement of a degree of arc of one minute of a meridian (i.e. as
an approximation of a nautical mile's measure), as well as the calibration
of the hour glass against a clock or sun dial, etc., at the time of its making. Which is to say, there was a great deal of knowledge and know-how
embedded in the tying of knots in a line. These knots might then be
distinguished by tassels indicating numbers, which could, in theory, be
identified at a glance as the rope ran over the stern into the water (assuming, of course, a basic level of numeracy).[36]

There was nothing particularly mysterious or prestigious about any
of this. Nevertheless it was crucial – and often difficult – work that had
to be done, and done well, by a team of sailors working in close coordination. Repeated trials were called for to compensate for human error,
unsteady hands, uncooperative sand, the give or shrinkage of rope, the
strength of a current, or the yaw of the ship. Whatever confidence might
have been inspired by Crignon's actions and cosmographical theories
was thus also tied to the hands on practice of La Pensée's crew's handling of logs, ropes, knots, and hourglasses. This, of course, was true not
only for the determination of the ship's speed and bearing, but for any
number of activities governing life on board a ship sailing the open seas.
Trust, in this sense, had to be distributed across an entire crew working
together as a unit to keep the ship afloat and on course. Work had to be
cooperative, an *esprit de corps* bound together by the exigencies of winds
and currents and harnessed to time: everyone had something to do or
a station to man at a specified hour. Put another way, the translation
of winds, currents and tired hungry sailors from potential enemies into
allies working together to steer ships across unknown seas, required a
great deal of coordinated work and discipline: rigging had to be checked,
capstans and deadeyes managed, sails had to be watched and maneuvered, bilge pumps had to be manned, caulk had to be applied, fish had
to be caught, and seas had to be surveyed for land and brigands. These
were activities directed by captains and pilots, not typically performed by
them.[37] The ship was, in this sense, a kind of collective body, animated by
the human bodies of a crew that performed on – and below – its decks to
the rhythm of a carefully maintained metrology of ship's watches and to
the commands given by those with "superior understanding" and social
authority, such as Crignon and Parmentier.[38]

[36] See, Hutchins, *Cognition*, 103–107.
[37] Ibid., 17.
[38] Ibid., 76–77. Watches were a crucial part of this regimen. Glasses of sand would be
manned at all times, usually in duplicate in case something went wrong with one of

The essence of any ritual is carefully enacted repetition. Whether taking the height of the sun each day at noon, determining a ship's speed, checking the rigging, operating the bilge pumps or turning glasses of sand, it was through the multiplication of trials and constant rehearsals that anomalies and variations introduced by human and instrumental error could be minimized and subsumed into a larger pattern. Crignon's attempt to garner authority and status and win confidence through his virtuosic and esoteric display of navigating the ship with astrolabe, compass, and chart thus extended to – and was integrated with – readings made by and through the bodies of the crew. Trust, in this sense, circulated in a kind of feedback loop through bodily and instrumental back-up systems that displaced responsibility while at the same time generalizing it. Risk and error, introduced by contrary winds, dangerous natives, unruly and error-prone men, diseased bodies and leaking hulls, could thus be spread out over multiple bodies, technologies, instruments, and readings – a kind of bodily, technical and representational equivalent of a joint stock company.[39] Confidence thus recapitulated scale – men and technology acting together to turn the tables on, co-opt and balance out, the threatening forces of nature and human and technical error. Like the Spaniard they met in the Cape Verdes, they too needed a company of well-armed allies that could present a coordinated front as a means of either overcoming hostile conditions, or minimizing the damage that they might cause.[40] In other words, one needed to compensate for doubt and suspicion with the concerted actions of men acting in concert, an *escale*, a regimentation that disciplined individual actors into an integrated and harmonious whole – a kind of collective body (a host)– as a means of doing what one man alone could not, thus creating a kind of balance of force, an equilibrium, between ship and the exigencies of the dangerous world being navigated, in the same way that the unnamed Spaniard on his remote African outpost balanced out (with the presence of allies) the (perceived) threat posed by Parmentier and his crew.

them. They were typically calibrated either to four hours or to half hours, but as we have also seen, to smaller times according to specialized uses such as determining a ship's speed. With each turning of a glass, bells would be rung so that all could hear. On Spanish vessels, these bells were accompanied by prayers; like a rosary, each flip of the glass was accompanied by a chant or prayer.

39 See in particular Mialet's *Hawking Incorporated*.

40 See, for example, Latour's essay, "Give me a Laboratory and I'll Raise the World," in Karin Knorr-Cetina and Michael Mulkay (eds.), *Science Observed: Perspectives on the Social Study of Science* (London, Beverly Hills and New Delhi, 1983), 141–170; Michel Callon, "Some elements of a sociology of translation: Domestification of the Scallops and Fishermen of St. Brieuc Bay," in J. Law (ed.), *Power, Action, Belief: A New Sociology of Knowledge? Sociological Review Monograph 32* (London, 1986); and Law's "On the Methods."

Malagasy Interludes

For 71 days they sailed. They crossed the equator on May 11; the journal notes that they sighted an unknown island, which Parmentier named France, after the French king, on May 29, but conditions did not permit them to approach. Around two months later they saw a great number of birds and guessed (*estimons*) that they were near the island of Saint-Laurent, also known as Madagascar.[41] On July 24, in the second quarter of the second watch, the wind suddenly changed directions. They saw that the seas were troubled and so sounded the bottom at six or seven brasses, and then, in the distance, they saw land. The next night, as they approached, great fires on shore were visible. Who had set them, were they friend or foe, helpful natives, dangerous savages, enemy Portuguese? Whatever the answer, after three months at sea, they had to find out.

The French were far from being the first Europeans to visit Madagascar;[42] the Portuguese had been coming for years. Their relations with the indigenous peoples were characterized by violence, deception, and rapacious greed. From first contact they kidnapped, stole, and indiscriminately murdered. In 1506, for example, Tristan da Cunha, with the ships of Jean Gomes d'Abreu, Jean Rodrigues Pereira and Job Queimado, made for Saint Laurent. They arrived near the Bay of Mahajamba on the island's northwest coast. The admiral sent two boats to keep watch over the strait and to stop the inhabitants of a small island in the bay from escaping to the mainland; he then went to the port and dropped anchor. The Moors, seized with fear sought to flee, some in their dugouts, others by swimming away. Many, especially the women and children, drowned. Their floating bodies covered the river. The admiral and his sailors went ashore and massacred those who were left and then sacked the village. This was not, the chronicler tells us, very profitable.

The Malagasy weren't simply powerless victims of European aggression however; they were more than up to fighting a war of mutual plunder. In these encounters they showed themselves capable of strategies far more devious than the simple smash-and-grab tactics of the Portuguese. In 1527, for example, only two years before the expedition from Dieppe

---

[41] Crignon, *Oeuvres*, 25.

[42] Parmentier's wasn't even the first French ship to land there. Two years previously, one of Verrazano's fleet was separated from the others while passing the Cape of Good Hope and landed there thinking it was India. They traded for all manner of goods, finding the natives welcoming and hospitable. They were perhaps too credulous and trusting, however, for when they arrived back in Dieppe they discovered that the merchandise they had brought back was all but worthless. See Gaspar Correa, *Lendas da India*, t. III, p. 241 as cited in A. Grandidier, Charles-Roux, Cl. Delhorbe, H. Froidevaux et G. Grandidier (eds.), *Collection des ouvrages anciens concernant Madagascar*, 8 tomes (Paris, 1903), I: 59.

set out, two Portuguese ships ran aground while pillaging the southwest coast of the island. Some of the survivors settled on a small island at the mouth of the Fanjahira river (now called the Efaho). The locals persuaded the Portuguese to hold a kind of "house warming" feast. Between 500 and 600 natives came. They implored the Portuguese to show them the wares they had salvaged from their ships so that they could marvel in wonder at them. After the feast, when the Portuguese were good and drunk, the Malagasy attacked. Those who survived retreated to their newly built fort with 30 or so native allies. They staged one raid after another, pillaging villages in the vicinity. The Malagasy king eventually sued for peace, promising to supply them with food if they would leave them be. In 1531, five survivors were rescued by Diogo da Fonseca. Da Fonesca also rescued a Frenchman. A survivor of Pierre Caunay's failed expedition, the Frenchman told them that so long as he possessed European wares he was mistreated and abused; but once he had lost everything, and went about naked like them, he was treated very well. It seems blending in and going native was a better inducement to trust than the possession of exotic goods.

Given this history, who was to say how the natives would react to the arrival of the French? Would they distinguish them from the Portuguese? How could they? Clearly, they would have to be convinced of their good intentions. This was not easy to do. Upon arriving they sighted four natives in a long canoe; one of La Pensée's boats approached, but before they could say or do anything, the Malagasy abandoned their *barquette* and swam away.[43] Le Sacre, however, sighted another, and followed in hot pursuit. They captured two Moors and brought them back to the ship. They were given bonnets, beads (*patenostres*) and cloth (*bourgran*) as tokens of friendship before being set free. Being chased, taken prisoner, and given gifts, was supposed to make clear the good intentions of the French. One can imagine that these confidence-building strategies might have had the opposite effect.

Deferring to the expertise of La Pensée's maistre, Michel Mery, and Le Sacre's captain, Raoul Parmentier, it was decided that the reefs were too great a threat to risk going ashore. Two brave and judicious men (*vaillans, bien deliberez*), Vassé and Jacques l'Escossois, however, asked leave to swim for it.[44] They were warmly welcomed by natives carrying fruit for sale. The next day, three or four Moors arrived by boat at Le Sacre carrying a goat and some fruit (La Pensée was farther out at sea); they were given bonnets, bougran, and patenostres in exchange. That evening the

[43] Crignon, *Oeuvres*, 25.
[44] Ibid.

French left, travelling north along the coast to look for a place where they could more easily make landfall to find water and supplies.

The next day the boats were sent out on a reconnaissance mission. They were ordered to return without exposing themselves (*sans s'exposer sur la terre*) because of previous, unspecified, but clearly bad, experiences with the Malagasy in the South.[45] The Moors here, Crignon says, were of good cheer (*chere*), making a show of putting their weapons away and sending them back to the forest with two boys. This confidence-inspiring measure was reciprocated by the French who were heartened (s'*enhardirent*) by this demonstrative act of welcome. They thus left their weapons on their boats and arrived instead with beads. By gesture and sign they were given to believe that if they followed the natives into the forest they would be led to a place where there was a great deal of ginger (*zingembre*), which they called "chellou," and also to forgers (*forgeurs*) of gold and silver.[46] Persuaded of the possibility of untold riches, the *contre-maistre* of Le Sacre, Breant, Jacques l'Escossois, and Vassé, along with a few others, followed the natives into the forest. Moments later, Jacques cried out, and Breant and Vassé burst from the tree line with 16 or 18 armed Moors in hot pursuit. The men on Le Sacre saw this and sounded the trumpets to warn those on shore filling barrels with water to get back to the ship immediately. Breant, Vassé and Jacques had been killed. The Moors chased the others to the water's edge. As they swam to safety, one of the Moors held up the bloody shirt of Jacques the Scotsman. According to Crignon, because the Malagasy couldn't "restrain the living among our men," he threw the bloody shirt to the ground and stomped on it (*pila dessus*). The Moors then looted the bodies of the dead, each taking a portion (*leur piece*) before returning to the shore to wash the blood off in the sea.[47]

According to Crignon, the shirt covered with the Scotsman's blood was for the Malagasy a metonym of the collective body of the French over whom they had triumphed. This identification, however, according to Crignon, was clearly the result of frustration; it was a kind of second-best trophy as the rest of the French were too far away either to capture or kill. Crignon's account of Malagasy treachery, however, hints at a more expansive frame of reference than the encounter alone, one that provides a glimpse into the mind of a man from a city and a profession that were being profoundly influenced by the new religion. One can perhaps discern in this description of the Moor and the bloody shirt the

[45] Ibid., 26.
[46] Ibid.
[47] Ibid.

doubts that were beginning to wend their way through the theology of presence in Northern Europe at the time of the Reformation. Crignon, after all, was here putting the metonymic power of a blood relic into the hands of a rapacious and untrustworthy Moor. Encounters with others could be unpredictable indeed.[48]

There were, of course, practical points of reference upon which cross-cultural encounters could turn. These involved the careful weighing of the strengths and weaknesses of strangers relative to one's own: how many were there? Were they armed? Did they have allies nearby? Or perhaps enemies? Did they have water, food, women, spice, or gold? Were they Christians or Pagans? Some of these questions could be answered with a glance; others required more substantial interactions. Once on shore, a new language had to be spoken and understood: one of gestures, facial expressions, movement of eyes, poise of bodies, tone of voices, and of course, the indexicality of the ostensive gesture of hands pointing – this is what we have here: knives, mirrors, beads, bonnets, and cloth, for what you have there: water, fruit, ginger, and gold (e.g. like the piece you're wearing around your neck which I'm touching with my finger). A sailor was experienced in reading nature's signs: observing the wake of his ship to estimate its speed, feeling the wind against his body to know its direction and power, sensing the motion of the yaw to feel the strength of a current. In all probability he was also used to working amongst a linguistically diverse crew that spoke vastly different languages and dialects: Norman, Breton, and Basque as well as Spanish, Portuguese, Scotch English, Flemish, and Italian, etc.[49] Sailors surely relied on specialized ship board pigeons of one kind or another to communicate, but also on languages of signs and gestures. Could they deploy a similar kind of hermeneutics in their encounters with others? How far could such a language take them in assaying the intentions and in taking stock of their relative strengths and weaknesses?

Given time, trust might be calibrated to conventions of sign, language, value, weight, and measure. Or it might simply be adjudicated by

---

[48] Perhaps even more so as there were no celestial bodies against which they could be reliably calibrated. There is no indication that Crignon's expertise extended to astrology, as it did for men like Fine, Gemma Frisius, Mercator, and Dee.

[49] As Charles de Bovelles said: "there exists in France at this time as many human customs and languages as peoples, regions and cities." Quoted in Paul Cohen, "Mediating Linguistic Difference in the Early Modern French Atlantic World: Linguistic Diversity in Old and New France." Working Paper in International Seminar on the History of the Atlantic World 1500–1800, Harvard University (2003); Colette Dumont-Demaizière, ed., and trans, *Sur les langues vulgaires et la variété de la langue française. Liber de differentia vulgarium linguarum et Gallici sermonis varietate* … (Paris, 1973), published originally by Robert Estienne in 1533, 5–6.

a balance of force: who was stronger, who was weaker, and bluffing or playing it safe when one wasn't sure. But time was of the essence, and this was not the Parmentiers' final destination. Risk defined the situation; anything could happen in a context without mutually recognized constraints. Why then did the French leave the relative safety of the beach and the company – and cover – provided by their compatriots and walk into the forest alone with the Malagasy? In other words, why did they cooperate with strangers? Why did they have faith in them? Why didn't they take precautions?

Perhaps it was less a question of trust in strangers than it was the confidence they had in their own abilities to protect themselves regardless of the good faith of the Moors. Perhaps they thought the balance was tipped in their direction in their dealings with these clearly inferior peoples. Whatever the case, they badly misjudged the situation. In this sense, the Frenchmen were entirely complicit in their own deaths; they went because they wanted to go; and they wanted to go because they weren't just sailors looking for water and essential supplies, but because they were also merchants who smelled money. Potential profit clearly outweighed the risk; indeed, why else would they have embarked on a mission of such magnitude in the first place? Unlike Parmentier, they were not sailing around the world for honor, king, or God; rather, they were risking their lives for profit. For them, it was clearly the case that the potential of great riches balanced out any risk that things might go wrong in their dealings with (clearly inferior) native trading "partners." The problem was that the criteria that they relied upon to assess the likelihood of profit had more to do with unrestrained desire reminiscent of Sallust's description of Lucius Sulla's men in Asia than with any rational assessment: in other words, they heard what they wanted to hear and saw what they wanted to see; they weren't thinking, thus they blindly followed the Malagasy into the forest. Perhaps they too had tricks up their sleeves; maybe one of them took a fancy to a trinket, or to one of the Malagasy women, or insulted – or spooked – them by making the wrong gesture or sound. Who knows what happened once they disappeared beyond the trees. But this is the point: they went either because they trusted the savages and/or because they trusted in their own capacities to defend themselves. Or more likely than not, they weren't thinking about their safety at all, but rather about potential profits and gain – the promise of spice and precious metal. In this sense, it was their unrestrained desire that got the better of them, coloring their perceptions of the encounter and clouding their judgment. The word they heard for ginger, for example, *chellou*, does not exist (in any variant, e.g. selo or tselo or sailo or tsailo) among the Sakalava and Vezo people who inhabited the region of Madagascar where they landed (rather, they call ginger sakarivo or sakaviro).[50] What else did they not see as they recklessly followed the natives into the forest looking for imagined ginger, gold, and silver? This is perhaps why Crignon used the word forgeur when he referred to native manufacturers of gold in his journal; at one and the same time, the word forgeur meant maker and counterfeiter, inventor and devisor, one who mints and fabricates precious metal into coin, and one who skillfully crafts the false, the base and the artificial and presents it as true.

The captain and the shipmasters were, according to Crignon, very angry (*fort courroucet et marris*) when they heard what had happened. Was their anger directed at their crew or the natives? In either case, their anger seems to have dissipated quickly, as Crignon's Log shifts suddenly from their reaction to the death of their men to the "taste of pepper in the air," and to what appeared to be an outdoor arena "sewn with little lights or scales of gold or silver, small like grains of sand."[51] The natives were clearly untrustworthy, but there was still ample reason to risk going to shore again. The next day, they snuck into the arena, finding there a grain or two of fine silver. They quickly returned from their surreptitious explorations. After saying a mass for their dead, they waited until day's end before returning to shore to find water and to see if there were really mines of silver or gold.

There is a discrepancy at this point in Crignon's journal. Perhaps he and the men wanted to look for the mines, as this is what he said they were going to do, but it seems that the Parmentier brothers had other ideas. Led by their captains, four boats, equipped with sailors and soldiers arrived on shore. There was no more talk of gold or silver; they immediately set off to find their fallen comrades. Breant was found first, buried just outside the forest in a shallow grave a half foot deep and covered with palm leaves, a large "boise seche" and with reeds. Just beyond the trees they found Jacques l'Escossois' nude body resting face down (*couché aux dents*); his chest was covered with arrow wounds and he already smelled very strong. They dug a grave and placed him inside. They found Vassé's nude body next, also lying face down, pierced transversally through his back and sides, his entrails plainly visible, his body riddled with arrow wounds. When they turned him over, his guts spilled from his stomach.[52] They dug another grave and prayed to God that he

---

[50] I would like to thank Jean-Marie de La Beaujardière, project director of the Malagasy Dictionary, http://malagasyword.org for this information.

[51] Crignon, *Oeuvres*, 27.

[52] That they were concealing the body on the beach as a means of covering their tracks becomes clear insofar as the others that they killed were left where they had fallen in the jungle.

take pity on their souls. This done, they went back to the spring in the north, and the men diligently, and in good order – as directed by their two captains – filled the ship and got underway. They then saw the arena on the edge of the sea and it seemed to glow with silver and it "was concluded… by those "who say of themselves that they know" (*se disorient a ce connoistre*) that this was indeed a mine of silver.[53] One can imagine the longing looks of the men viewing the temptations of this dangerous shore. Was a simple self-referential claim (those "who say of themselves that they know"), enough to warrant trust – confidence in expert knowledge – that they would indeed find silver and silver enough to offset the obvious risk? Was their assessment of the potential for profit creditable? For the captains, surely not; in their view the time and the cost (*le temps et le coust*) it would take to have a sufficient quantity of silver to make it worthwhile was lacking; it was decided, therefore, that they had more to lose than to gain by returning to shore.[54] The Parmentiers' authority thus triumphed over rival claims of expertise, and served to moderate, or at the very least, refocus, the crew's greed. This was not, however, the last time that differences of opinion and clashes of expertise were to appear in Crignon's matter-of-fact log.

Given their reception, "it was concluded among our captains and masters to leave this place with the first good wind."[55] This decision was immediately validated when they noticed that their movements along the coast were being shadowed by a troop of Moors armed with the same kind of weapons used to kill their compatriots the day before. Signs were made to them, but they were not understood, nor, says Crignon, did they understand Portuguese. The French shot at them, but they were not fazed; this led the French to believe (*estimons*) that these Moors weren't familiar with artillery. They learned quickly, however, that the French guns could be deadly when the accurate eye of the "Flemish" from Le Sacre found its target. Clearly the French were now "standing their ground."[56]

Beyond the Islands of Fear: Where Am I Redux

On Saturday, July 27, at noon, about two leagues from land, the height of the sun was taken at exactly (*justes*) 19 degrees, according to

---

[53] Crignon, *Oeuvres*, 27. My emphasis.
[54] Ibid.
[55] Ibid.
[56] Ibid., 28. There is a curious detail in Crignon's description as he notes the presence of a "white Moor." That this might have been a castaway from a previous expedition is a possibility; that Crignon believed that they did not understand the threat posed by guns casts doubt on this supposition however.

the "declination" [calculated] by the Portuguese. According to Pierre Mauclerc, however, who was the astrologue of the Sacre, they were at 25 degrees and 52 minutes. The disagreement was substantial, but remained unresolved. They proceeded carefully, regardless of where they were in an abstract system of mathematical coordinates, making their way through what they called the Islands of Fear (*les isles de Crainte*). The seabed below was treacherous and uneven; confidence threatened, they deployed one of the small boats to sound the way before them. On August 3 the waters became rough, indeed, angry (*fascheuse*); they called them the Sea without Reason (*la mer Sans Raison*). Moreover, since passing the cape of Good Hope, the men had become weary, unreliable and frustrated (*las, faillis et vains*); many were suffering from severe back pain, high fevers, and ulcerated sores over thighs and legs.[57] The following day the height was taken at noon and was found to be 17.5 degrees. On August 8 they arrived in a chain of islands.[58] They sent a boat out to see if there was a place where they could safely approach. They saw a village with "500 hundred people" who came toward them; they were large men wearing clothes.[59] They traded with the French: knives and bonnets for coconuts. When the natives saw how "liberal" (*liberaux*) the French were with their goods, they demanded a high price for their spices (their *drogues*). This seems to have put Parmentier and his brother on edge. Though they had begun filling barrels with fresh water, when they saw other savages (*autres sauvages*) descending from the mountains, they quickly returned to their ships. They put to sail on August 11. The next day Pontillon died and Parmentier performed his anatomy (*le capitaine fit faire une anatomie*); the day after he performed a postmortem on Jean Dresaulx. His display of expertise was perhaps meant to inspire confidence, i.e. that his esoteric knowledge of the blood and guts of his men might provide him with some kind of preternatural insight into their predicament and the best way out of it. One can imagine that it had a very different effect on the crew, notwithstanding the captain's obvious learning and piety. They continued, carefully, on the Sea without Reason.

A month later to the day, on September 13, Pierre Le Conte died. He had not eaten for more than three weeks. A few days afterward, a Sannais sailor, "a good man" (*bon homme*) named Guillemin Le Page who had long suffered from back, leg, and stomach pain, died.[60] Fortunately, on September 20 they sighted land again, a group of islands that they said

---

[57] Ibid., 29. These symptoms are consistent with scurvy.
[58] According to Nothnagle, this was probably the Isle of Mayotte in the Comoros. See ibid., 51, n. 25.
[59] Ibid., 30.
[60] Ibid., 33.

were beyond Calicut to the north and the Comoros to the south. They found a place to go ashore on a green island planted with many palms.

Jean Masson, the translator (*notre truchement*), led two of the small boats to shore. They were welcomed by the natives with palm fruit and figs. Masson gave them two knives, some mirrors and some trinkets in return; a gift was brought for their captain, a small and ingenious chariot made from a single piece of wood that folded in two, and also two or three pounds (livres) of palm sugar, which they called Zagre and around a quarteron (a quarter of a pound) or a "demi cent" of unrefined black sugar. The next day, another sailor, Jean François, died, and Parmentier – "himself, in person" – went to shore, accompanied by the two large boats filled with well-armed and disciplined men. He was *"fort honnestement"* received by the principal and great archpriest (*archiprestre*) of the island who kneeled before him and kissed his hands. The archrpiest then presented Parmentier with a big lemon, round like a large orange. The captain rushed (*courut*) to have him rise and embraced him. He then presented him with two large knives. The native workers, the *menu peuple* of the island, then climbed palm trees and brought them coconuts, which they broke for the Frenchmen so that they could drink the water inside.[61]

Crignon recounts his impression of their magnificently made temple or mosquette and the intricate carvings that covered it inside and out. These were, he says, the most beautiful the captain had ever seen (*les plus belles qu'il vit jamais*). At the end of the structure, he describes a "secret" place, a kind of "*Sanctum Sanctorum*," that Parmentier wanted to open. He was curious to see what was inside and to know if they had any idols, but he saw only a lamp made of coconut shells.[62] There was nothing particularly special about their homes, he continues: they were small and ordinary. The people too, he says, were small, and they were also very thin. He notes also that there were no young women anywhere to be found: only "old, skinny, poor and grey" ones (*vieilles, maigres, pauvres et chenues*). Moreover, they had practically nothing in their homes, or at least, nothing of worth. Crignon took this to mean that they had moved everything of value, including their women and children, to a safe place toward the interior of the island, "fearing that we would take them by force."[63] This, he says, was the work of the archpriest who, he says, was "a great and wise man" (*un grand sage homme*).[64] Crignon thus praises the native priest for the perspicacious knowing, or at least suspecting, that

---

[61] Ibid., 34.
[62] Ibid., 35.
[63] Ibid.
[64] Ibid.

the French were not to be trusted. This seems odd at first glance. The archpriest's mistrust of the French – or at least the precautions taken in advance of the arrival of strangers – was here taken as evidence of good judgment, prudence, and trustworthiness. Crignon, however, reveals the true motivations for his praise in the very next sentence, where he introduces a "little disagreement" (*un petit estrif*) between Parmentier and the "Portuguese" from Le Sacre.

The Portuguese, says Crignon, "wrongly" told his companions that they were on one of the Maldive islands. Crignon explains, they were actually half a degree south, and that the Maldives were between seven to seventeen degrees north of the line. Though the Captain pointed this out to him, the Portuguese persisted in his erroneous position, and even tried to enlist the archpriest to his cause. In other words, not only did the Portuguese disagree with his captain on fundamental matters of navigation (that is, in knowing where they were), he relied on the credibility of local knowledge to sustain his objections. That he considered the archpriest a potential ally in his disagreement with his captain is intriguing; it challenges assumed asymmetries in indigenous European encounters, and reveals the deference that was often accorded to the knowledge of native informants, whether of weather patterns, edible or medicinal plants, the locations of enemies or possible trading ports.[65] The Portuguese's confidence was no doubt based on the fact that he took it for granted that the archpriest (who "was a knowledgeable man who had seen much")[66] possessed detailed and accurate information regarding the region where he lived. For these reasons, the captain and Crignon also trusted him. Their trust, however, was also based on other creditworthy qualities enumerated by Crignon: e.g. that he wore clothes, that he was not a worshipper of idols, and that his people possessed a highly developed material culture. Crignon's description thus had the narrative effect of setting up the "knowledgeable, pious, humble, and friendly" archpriest as a trustworthy arbiter between Parmentier and the "Portuguese." In Crignon's reconstruction, the now "creditable"

---

[65] On the colonial "extraction" of indigenous knowledge see, for example, Laurelyn Whitt, *Science, Colnialism, and Indigenous Peoples: The Cultural Politics of Law and Knowledge* (Cambridge, 2009). Timothy Walker, "Acquisition and Circulation of Medical Knowledge Within the Early Modern Portuguese Colonial Empire," in Daniela Bleichmar, Paula De Vos, Kristin Huffine, Kevin Sheehan (eds.), *Science in the Spanish and Portuguese Empires, 1500–1800* (Stanford, 2008), and Michael Bravo's "Ethnographic Navigation and the Geographical Gift," in David N. Livingstone, Charles W. J. Withers (eds.), *Geography and Enlightenment* (Chicago, 1999), 199–235; on the role of go-betweens, see for example, Kapil Raj, "Go-Betweens, Travelers, and Cultural Translators," in B. Lightman (ed.), *A Companion to the History of Science* (Chichester, 2016), 39–57.

[66] Crignon, *Oeuvres*, 35.

archpriest was recruited into the conflict on Parmentier's side, to affirm that "this island has the name Moluque [probably the Mulaku Atoll], and that the islands of the Maldives were at least two hundred leagues to the north."[67]

Perhaps the archpriest was putting his finger to the wind and tactically siding with Parmentier as the man clearly in charge of potentially dangerous foreigners. It is possible that their discussion took place with maps and charts, or with fingers drawing in the sand. No further details are provided. But given the limits imposed by linguistic and cultural difference, these interactions must have been more confusing than comprehensible. The archpriest's defense of Parmentier's position is left to stand on its own, but then, suddenly, and for the first and only time, the future intrudes into Crignon's narrative in the form of an addition where he notes that "notwithstanding" this disagreement and its subsequent resolution, he has "since seen a Portuguese map where the islands below the line are called the Maldives."[68] In other words, the opinion of the Portuguese is borne out; he was right and the captain was wrong: the Maldives were not where he thought they were, and neither were they. The intrusion of the future into Crignon's Log reveals its editing and correction at a later date, at a time when the captain was not in need of support and validation. This left Crignon, his trusted adviser, free to correct Parmentier's erroneous opinion, while affirming his (Crignon's) own continuous engagement with – and expertise in – questions of navigation and cartography.

Rather than return to the "little disagreement," between the Captain and Antoine, the Portuguese, Crignon returns to a discussion of useful local knowledge – to the archpriest showing the captain the wind lines for various parts of the world including the "lands of Adam, of Perse, of Ormus, of Calicut, of Zeilan, of Moluque and of [their destination] Sumatra." The Captain having got what he came for – knowledge and food – paid what he considered a fair price and loaded the boats with chickens and long green figs. Before getting underway, Crignon looks back one more time at the village, "they call their God Allah," he said.[69] Perhaps this final observation, almost an afterthought, was meant to cast doubt on the archpriest's credibility: his religious confusion being conflated by Crignon with his geographical befuddlement, as revealed by an after-the-fact viewing of an anonymous Portuguese map. That night, after eating, they set sail: south by south east.

---

[67] Ibid.
[68] Ibid.
[69] Ibid., 36.

Ticou France et France Ticou

The next day, the height of the sun was taken, two-thirds of a degree to the south; a week later Crignon estimated that they were 55 leagues from Taprobane, that is, from their destination in Sumatra. On October 6 Aleaume de Rambures, after having been very sick, had a hard death. On October 29 Crignon reports a possible eclipse; it was an overcast and rainy day; no mention is made of longitude as one might expect, but rather the death of a trumpeter, Beausseron, is noted. Two days later, a sailor on La Pensée sighted land; they made their way through a series of small wooded islands. Another sailor, one of the pages, died. His name was Barbier. Off the coast of one unnamed island they found several well-made traps filled with fish of many "colors and shapes," but when they went ashore they found no one. On Saturday and Sunday Masses were sung by their chaplain, and on Sunday, while they were on their way to land, Le Four, the son of a "wafer-maker" named Thomassin le Boulanger, died. On Wednesday, the eve of Saint Simon and Saint Jude, they made their way through islands they named Parmentier, after their captains; Marguerite, after the king's sister; and Louise, after his mother. A locksmith named Nicolas Boucher died, and the next day, a silversmith named Colinet Fayolle. On Thursday they sighted a large land mass; three days later, on All Saint's Day, they sent the big boats ashore, where they were greeted by 30 well-armed men.

Jean Masson spoke with them, and they approached with rice in exchange for mirrors and knives, and a rooster and chicken, which they sent back, because the natives wanted too much – a bougran – for them.[70] They showed the French some pepper and told them that there was a great deal of it on the island and also a lot of gold. The inhabitants of Sumatra's west coast were used to strangers coming to trade. They surely had the skills to create an ad hoc pigeon of Malay, Minangkabau (a patois of Riau Malay), Batak, the Min Nan dialect of Chinese, possibly Arabic, and certainly gestures, signs and props. Indeed, as Anthony Reid points out, "linguistic diversity was part of the everyday experience of Southeast Asian commerce."[71] Where Jean Masson, their translator, learned Malay is not mentioned; it is most likely from having served on a Portuguese vessel.[72] In any case, his abilities were surely rudimentary. It was clear, however, that they were being directed to a place where they

---

[70] Ibid., 39.

[71] Anthony Reid, *Charting the Shape of Early Modern Southeast Asia* (Singapore, 2000), 159.

[72] The learning of languages in the Indes was a priority for the Portuguese. See Cláudio Costa Pinheiro, "Words of Conquest: Portuguese Colonial Experiences and the Conquest of Epistemological Territories" *Indian Historical Review* (2009): 37–53.

could carry out more substantial trade – to the South in a village called Ticou ruled by a king they understood to be named Sultan Megilica.[73] To show them the way they took three of the natives on board, giving them each a bougran rouge, five quarters of white cloth, a knife, and a mirror in exchange for their knowledge.

On Tuesday morning, the "Day of the Dead," a skiff arrived with three emissaries from Ticou asking that the captain make haste to see the king as he wanted to present him with a gift. Parmentier responded politely that he would, in time, thank the king, and that he would also make an "*honneste*" gift, and go see him. They hauled anchor and set off, and when they arrived off shore of Ticou another skiff arrived with various gifts from the king, including two goats, some chickens, a bushel of rice, betel leaves steeped in hot water, and a finely minced root in a copper dish. The betel, Crignon explains, "is often eaten by the natives before or after a meal; it smells good, and it has a juice that stains their teeth red."[74]

The next day, another skiff arrived to tell them that the king understood that Parmentier planned to come ashore. The captain, however, replied that he wasn't going anywhere until he had "*pleiges*" on board the ship. Stated simply, Parmentier wasn't given to trust; he needed substantial assurances if he were to leave the safety of La Pensée. A balance had to be struck and a guarantee given.[75] Trust, in this sense, was embedded in a material exchange of bodies, bodies that forfeited – or at least, deferred – humanity to become surety-objects to be weighed against "objects" of commensurate value given by the French. The mutual exchange of pledges was meant to create a state of equilibrium – a balance of forces that could act as a standard against which subsequent interactions could be regularized and peacefully maintained.[76]

Pledges, of course, straddled many potential roles: credit instruments (i.e. instruments of surety), ambassadors, and spies being only the most obvious. Indeed, a pledge could serve as a subtle, perhaps clandestine, means to learn about strangers: their language, their customs, their food, their numbers, their friends, their enemies, their weapons, and their "true" intentions. The Sumatrans replied that they would have their pledges, and it was decided that Jean Masson, Nicolas Bout and Crignon would go to shore "to have knowledge of those of the land" (*à celle fin*

---

[73] Crignon, *Oeuvres*, 39. Perhaps a reference to a provincial governor (*mendelika*) rather than a sultan.

[74] Ibid.

[75] See Adam Kosto, *Hostages in the Middle Ages* (Oxford, 2012).

[76] Hostages need to be given, not taken. If the latter were the case they would be prisoners. Prisoners might be held hostage to ensure compliance, but this form of asymmetrical social interaction would be highly unstable and volatile.

que nous eussions connoissance de ceux de la terre). This, for Parmentier, was plenty reason to give them as *his* pledges.[77] Though ownership of his men would shift to the Minangkabau, their loyalty remained to Parmentier, who awaited their ultimate redemption, and the likelihood that they would collect valuable intelligence as a kind of surplus – an interest accrued – that could either help maintain a workable equilibrium through increased cross-cultural understanding, or perhaps more likely, help the captain tip the scales to his own advantage in their future interactions. As instruments these body-objects recall Parmentier's dissections of Pontillon, Jr. and Dresaulx: they were sites of inscription – "probes" – that could register and bring back hard-to-obtain intelligence that could be put to work to further the goal of a profitable voyage.

When the French pledges arrived they were courteously received and taken to the king's lieutenant, Tue Biquier Raza, who was waiting for them with all the most important people of Ticou (*avec toute la seigneurie de la ville de Ticou*). The pledges greeted him, thinking he was the king, as this is what they had been led to believe. It was however clearly beneath the king's dignity to meet with any but their captain. After salutations were made, they were brought to a place under a tree where Raza seated himself with his legs crossed "like a couturier" and gestured for the pledges to do the same. They were given betel to chew. The offering of betel was a "vital element in social relations and ritual transactions" among the Minangkabau.[78] It was more than just a mild and invigorating high – a social lubricant – to be employed in commercial negotiations; it was, as Anthony Reid puts it, "the essence of courtesy and hospitality."[79] Having accepted this opening gesture, the king's lieutenant proceeded to ask them who they were and why they had come. "Jean Masson "told him in the Malay language" (*luy dit en la langue malaye*) that they were French and that it had been 8 months since they left their country to come to see them and to bring them good merchandise…, and also to have their pepper." The Lieutenant happily listened to Masson's words.[80] Perhaps he leaned in to ask his next question: "were they not men of war?" This must have been the impression that the sea-hardened, well-armed, overdressed, round-eyed and hairy men of La Pensée and Le Sacre wanted to give to the coastal Minangkabau people of Ticou.[81] The encounter was no doubt tense. Masson responded that though "they were merchants

---

[77] Crignon, *Oeuvres*, 39.
[78] Anthony Reid, *Southeast Asia in the Age of Commerce, 1450–1680: The Lands Below the Winds*, Vol. 1 (New Haven, 1988), 5.
[79] Ibid., 44.
[80] Crignon, *Oeuvres*, 40.
[81] See Reid, *Charting the Shape*, 163.

and only asked for peace and love," for those who wanted to do them wrong "they were also men who could defend and avenge themselves."[82] Once this threat had been delivered, however, it was just as quickly softened with soothing words, a message from their captain, revealing his great desire to find "peace and love with them."[83] The king's lieutenant replied that they only wanted the same. Much else was said that Crignon could not understand as Mason, he complained, did not translate it back into French. They were then taken to the house of one of Ticou's principal governors and treated, "according to the customs of the land," to a delicious meal of rice and chicken that they ate with bread that they had brought with them from the ship. Afterwards they went to sleep using their coats as pillows. "God knows," Crignon recounts, "how we tossed and turned that night."[84] When morning finally came they returned to the ship accompanied by the Chabandaire, the "Harbor Master" of Ticou.

The Captain and the Harbor Master

The Chabandaire was a powerful official who governed everything having to do with trade and commerce in the name of his king. Not only did he set the prices for all the merchandise bought and sold in Ticou, he also established the weights and measures that would be used to adjudicate these transactions. "No one dare buy or sell without his leave," Crignon reports.[85] "God knows what beautiful promises this Chabandaire made to us," as much for selling their merchandise as for filling their ships with pepper.[86] He was received by the captain as if he were the king himself – showing him the various merchandise he had on offer. But the Chabandaire made it clear that until the king received his gift, there would be no trading. He demanded that this be taken care of immediately; indeed, that very day. Wanting to prepare something special, and to present themselves in good order in "*habits triomphans,*" the captain convinced the harbor master to wait until Sunday, November 7 (four days later) for the presentation of his gift.

To triumph was something with which Parmentier and his crew were familiar; triumphs were not just literary artifacts, written about by Sallust, Petrarch, or members of the Puy de palinod (such as Pierre Crignon's chant royal describing the Virgin's triumphant conquest of Satan and her

---

[82] Crignon, *Oeuvres*, 40.
[83] Ibid.
[84] Ibid.
[85] Ibid.
[86] Ibid.

redemption of all that Adam had lost), they were commonplace ritual performances enacted for the royal entries of kings, visiting dignitaries, for the arrival of relics, and for Christ himself in Corpus Christi day celebrations (see Chapter 1).[87] His crew would all have seen or participated in them. It was an interesting, if not provocative, choice for Parmentier to style himself as a triumphator (with *habits triomphans*) upon making his first official visit to Ticou. The triumphator was ritually defined as the instantiation of God's presence on earth – as he who comes in the name of the Lord (*Benedictus qui venit in nomine Domini*).[88] On the one hand, this presence demarcated a symbolic field of radical asymmetry, of conquest and submission, king and subject; on the other, it signified the opening bid in a prolonged series of negotiations over the range and limits of a triumphator's actual power.[89] In either case, one can imagine that Parmentier's fundamental goal was to tip the balance in his own favor in any future interactions with the Ticouans. Beyond the surely untranslatable cultural vocabulary of triumph that delayed his first meeting with the Minangkabau king, Parmentier's pretensions were meant to impress and intimidate. This said, his "well-ordered" and triumphant arrival was certainly directed as much at his own men as at the Minangkabau, who were surely less conversant than the French with the significance of the triumph. Enlisted to become a kind of ceremonial body with Parmentier at its head, the triumphant arrival of the French elaborated a ritual distribution of the captain's authority across the "well-ordered" ranks of his men's coordinated (military) procession. He, their leader, was the measure and the scale by which their (good) order was to be set out and performed; they in turn, were to become a corporate body (a host) whose unity of disposition and actions reflected his will and his desire. This, as we have seen, is the essence of the power of scale.

A Balanced Thought

Our view of what happened next is highly abbreviated, as Crignon was not able to attend because of an injury to his leg. The gift, he says, was honorably given and honorably received. Peace, friendship, and love were faithfully promised. The French swore to be "friends of the Ticouan's friends and enemies of their enemies. The natives reciprocated, and

[87] On the imagery and importance of triumph in early modern France see Wintroub, *A Savage Mirror*, chapter 4, "The Social Poetics of Triumph."

[88] See, for example, E. H., Kantorowicz, "The 'King's Advent' and the Enigmatic Panels in the Doors of Santa Sabina," *Arts Bulletin* 26:4 (1944): 207–231.

[89] See Lawrence Bryant's *The King and the City in the Parisian Royal Entry Ceremony: Politics, Ritual, and Art in the Renaissance* (Geneva, 1986).

they all exclaimed together, *Ticou France*, et *France Ticou*."[90] After this, Raoul and Jean arranged to rent (*louerent*) a house to store their merchandise. Jean Parmentier then spent the next eight days securing (*fortifier*) the house, and negotiating with the Chabandaire in setting weights, measures, and port fees. No mention is made of any translator helping in these difficult negotiations. Were they conducted with gestures and objects – with various kinds of weights and scales; or was the presence of translators simply effaced as a technical supplement meant to be invisible? Whatever the case, the negotiations proved difficult. The captain, says Crignon, found himself blocked at every turn (*fort empesché*); indeed, the two men found it difficult to agree about anything (*grand difficulté accorderent-ils ensemble*).[91] Parmentier perhaps believed in his own status as triumphator – as a civilizing agent of Christianity come face to face with a haughty savage. The Chabandaire, on the other hand, certainly would not have viewed Parmentier as a superior in any sense.[92] The Minangkabaou were experienced cosmopolitan traders.[93] The trust won with the exchange of pledges quickly dissipated, as their arguments over standards and fees escalated.

Beyond Parmentier's failure to understand the status of his interlocutors, agreement was no doubt difficult to achieve because of the very different forms of skill and expertise employed by each side in the work of standardization. As we have seen, even in France, weights and measures, coins and taxes, varied from city to city and town to town. The transvaluation of standards was difficult to achieve under the best of circumstances; in faraway places, among peoples with widely different cultures, values, expectations, and languages, as well as radically different skills,

[90] Crignon, *Oeuvres*, 41.

[91] Ibid.

[92] Reid, for example, notes that the chronicle of Malay kings, the Sejarah Melayu, calls the Portuguese "white Bengalis." According to Tomé Pires, "When they want to insult a man, they call him a Bengali," who were considered to be "sharp-witted but treacherous." See Reid, *Charting the Shape*, 162–163. This said, the French, like the Portuguese, were respected and feared for their military capacities; they could be powerful allies, or dangerous enemies. In either case, trade could be profitable and was generally welcomed. Ibid., 159 and 164–166. Pierre Caunay's reception in Aceh, however, stands as a stark reminder that this was not always the case.

[93] Their kings were widely treated with reverence and awe. Marsden, writing in the eighteenth century, for example, describes how the Batak of Northern Sumatra viewed the Minangkabau kings: "Notwithstanding [their] independent spirit ... and their contempt for all power that would affect a superiority over their little societies, they have a superstitious veneration for the Sultan of Menangabau, and show blind submission to his relations and emissaries, real or pretended, when such appear among them for the purpose of levying contributions." See Marsden, the *History of Sumatra* (Oxford, 1975), 337, and Jane Drakard, *A Malay Frontier: Unity and Duality in a Sumatran Kingdom* (Ithaca, 1990), 6. See also Drakard, *A Kingdom of Words: Language and Power in Sumatra* (Oxford, 1999), 44.

technical practices and forms of expertise, it must have been next to impossible, short of coercion.[94] William Marsden, for example, provides a detailed description of the metrological practices of the Sumatrans in the eighteenth century that perhaps indicate the nuanced level of expertise possessed by the Minangkabau in the sixteenth.[95] Before gold dust was weighed for sale, Marsden explains, "it had to be cleaned from all impurities and heterogeneous mixtures, whether natural or fraudulent (such as filings of copper or of iron)."[96] This was to be done by "a skillful person" whose "sharpness" of eye, and "long practice" enabled him to accomplish this "to a surprising degree of nicety." "The dust," Marsden says,

is spread out on a kind of wooden platter, and the base particles (*lanchong*) are touched out from the mass and put aside one by one, with an instrument, if such it may be termed, made of cotton cloth rolled up to a point. If the honesty of these gold-cleaners can be depended upon, their dexterity is almost infallible; and as some check upon the former, it is usual to pour the contents of each parcel when thus cleaned, into a vessel of aqua-fortis, which puts their accuracy to the test. ... [E]very man carries small scales about him, and purchases are made with it so low as to the weight of a grain or two of padi (rice). Various seeds are used as gold weights, but more especially these two: the one called rakat or saga-timbangan...being the well-known scarlet pea with a black spot; twenty-four of which constitute a mas, and sixteen mas a tail; the other called saga-puhn and kondori batang, a scarlet or rather coral bean, much larger than the former, and without the black spot.[97]

Marsden's description of expert knowledge, keen eyes, finely made tools, dexterous hands, backed up by assays made with specialized tools and techniques, surely demonstrates a level of sophistication regarding the determination of weights and measures on par with, or perhaps

---

[94] See, for example, William Pietz, "The Problem of the Fetish –1," *Res* 9 (1985): 5–17; Pietz, "The Problem of the Fetish –2," *Res* 13 (1987): 23–45; Pietz, "The Problem of the Fetish –3a," *Res* 16 (1988): 105–123. See also, Simon Schaffer, "Les cérémonies de la mesure: Repenser l'histoire mondiale des sciences," *Annales. Histoire, Sciences Sociales* 2 (2015): 409–435 and Wintroub, "Translations."

[95] Their metrological practices were based on longstanding customs and relations with the natural world – e.g. the usage of what were perceived to be naturally given measures based on commonly used grains, seeds, and plants (e.g. the *cupak* which was based on equal halves of a coconut shell or the length of a section of bamboo, or the weight of a saga seed which was a unit of measurement [a saga] used to weigh gold) – that changed only very slowly. See Yusharina Yusof, Zulkifli Ab Ghani Hilmi, and Saharani Abdul Rashid, "Weight Measurements in the Malay Minangkabau Culture," 2010 International Conference on Science and Social Research (CSSR 2010), December 5– 7, 2010, Kuala Lumpur, Malaysia, 861–866.

[96] William Marsden, *The History of Sumatra, Containing an Account of the Government, Laws, Customs and Manners of the Native Inhabitants with a Description of the Natural Productions, and a Relation of the Ancient Political State of that Island* (London, 1811), 170.

[97] Ibid., 170–171.

superior to, that of Europeans.[98] That these were not simply eighteenth-century practices is evinced by Tomé Pires' description of the region's complicated metrologies of measure, volume, and mass in the early sixteenth century.[99] According to the account of the "The Great Ship Captain from Dieppe," of the Parmentiers' voyage, often attributed to Crignon (see Chapter 6): "the people of this land [Ticou] do not use money, at least that which is brought from another country, they buy and sell everything according to its weight in gold; they measure cloth with a measure which is long as a cubit (a *coudée* – a length that extends from the tip of the middle finger to the elbow). They measure rice and pepper with a notched cane or reed, which contains about two pounds of pepper."[100] All of which is to say, the traders of Ticou were no pushovers; they were neither savages nor hicks, rather they were experienced and seasoned commercial experts who had highly developed metrological methods for adjudicating questions of weight, measure, and value. In his previous experience Parmentier had traded with naked "savages" in what is now Brazil; nothing had prepared him for negotiations with sophisticated and worldly traders such as the Minangkabau, except perhaps, his dealing with his own compatriots back at home. As Crignon said of them: "in commerce (*merchandise*) they are great bargainers (*grands barguigneux*), even more so than the Scots and the Dutch."[101]

The inability of the captain and the Chabandaire to find common ground reverberated far beyond the little rented house, leading from the subversion of the initial act of contact – and their calibration of pledges and gifts – to the destabilization of their encounter altogether. Realizing that they could never agree, the captains decided to return to the ships, taking their merchandise with them. The Minangkabau placed an extremely high priority on consensus (*mufaka*), being especially concerned with maintaining the work – the discussion, debate, the back and forth negotiations – which contributed to it.[102] For them, the process of deliberation (*musyawarah*) was as important as its end result. For Parmentier, the inability to achieve consensus with regard to standards and taxes made it impossible for him to trust the Minangkabau, hence

---

[98] "Based on a comparison of the magnitude of each measurement, a traditional Malay Minangkabau system of weight measurement for valuable metals is arranged as follows: *Baras/Beras < Uang < Saga/Sago = Kundi < Rakit = Kupang < Ameh/Emas < Pau/Pao/Paha < Tahil < Kati < Bahara.*" Yusof, "Weight Measurements," 863.

[99] See Armando Cortesao (ed.), *The Suma Oriental of Tome Pires. Account of the East, from the Red Sea to China, written in Malacca and India in 1512–1515.* 2 Vols (New Delhi, 2005), II: 275–278.

[100] Crignon, *Oeuvres*, 110.

[101] Ibid., 45.

[102] Drakard, *A Kingdom of Words*, 49.

his decision to unilaterally break off negotiations. For the Minangkabau, the opposite was the case, disagreement was simply a spur to continued negotiations – to more talk. They could not understand why Parmentier had just stopped talking with them. The Chabandaire grew increasingly distressed; his brother and another Ticouan were still being held as hostages on Le Sacre. Seeing that the French were preparing to leave, he assembled some 500 well-armed men to confront them. Antoine the Portuguese from Le Sacre begged the captain to immediately free the hostages (to avoid *tous belliqueux debats*).[103] He volunteered to remain on shore to serve for one (*demeurer pour un*); Crignon offered to serve as another.[104] The captain refused to part with Crignon. He chose Jean Le Peintre instead. They went to see the Chabandaire. When Antoine, the Portuguese, informed him what they had decided, he seemed relieved that the French were willing to once again recalibrate the balance of hostages – to talk again. However, he wasn't entirely satisfied. Jean Le Peintre, he felt, was not of sufficient weight or quality. He wanted Crignon. The captain resisted. Crignon argued that they had little choice given the instability of the situation. Parmentier relented and Crignon and the Portuguese left with the Chabandaire as his hostages. The rest of the French remained on shore with guns, swords, and shields at the ready. As the newly minted hostages passed through the village they saw "great waves" of well-armed Ticouans. Crignon feigned laughter (*faisois semblant de rire*); their host (*hoste*) looked at him and laughed too.[105] Was this bravado, was it fear, or was it a means of breaking through the palpable tension? The captain sent a boat for the "pledges and hostages," and waited. With the beating of drums and the sounding of pipes and trumpets, he ordered his men into battle formation. The Ticouans watched from a distance; they were, says Crignon, very afraid. A short while later, the Chabandaire came to Crignon and told him to deliver a message to the captain: they would not be released until the hostages had been returned. Crignon confided to Antoine that the situation looked grave; he feared they would be kept even after the Ticouan hostages had been released. Antoine reassured him, saying he had no doubt that the Chabandaire would keep his promise. Crignon, trusting in his shipmate's judgment, went to Parmentier and gave him the message. Parmentier was credulous. Regardless of what his men (the Ticouan's hostages) thought, he was suspicious. The captain ordered Crignon to stay and not to go back, but Crignon once again persuaded him that they had

[103] Crignon, *Oeuvres*, 42.
[104] Ibid.
[105] Ibid.

little choice but to trust that the Chabandaire would keep his word. The captain relented, promising Crignon that he wouldn't leave shore until both he and Antoine had been freed. When Crignon returned, the harbor master was puzzled; why, he asked, hadn't Parmentier simply left? Because, Crignon responded, he had given his word that he wouldn't.[106]

For the Minangkabau, oaths were thought to have an almost sacred power.[107] Jane Drakard has argued that the sacred character of the oath in Minangkabau culture was rooted in the identification of words and things, which she compares to the representational regime – "the episteme" – that characterized medieval Europe, and in particular, the doctrine of real presence associated with the Eucharist. "What is suggested," she says, "is that language, and particularly royal and religious language, had, in certain contexts, a substance in … Sumatran thinking which was akin to the solidity and evidential truth of material signs or *tanda*. Royal words, therefore could be regarded as 'marks of truth' encapsulating and embodying the word of God and be themselves evidential *tanda* of signs of God."[108] The Chabandaire was a spokesperson for the king. Parmentier was thought to hold a similar position, perhaps by the Minangkabau, but certainly by himself: i.e. as a triumphator who speaks for the absent God or king. The Chabandaire was greatly relieved to learn that Parmentier was a man who honored his word. But still, he wondered why it appeared that Parmentier was preparing for war? Crignon told him that this was because they had been warned that the Ticouans were out for their blood. The Chabandaire exclaimed that they had thought the same thing about the French. And then when the hostages were finally set free, they all embraced and once again sang their symmetrical song: "Ticou France, and France Ticou!"[109]

Between Balance and Incommensurability

The shift in terminology during these negotiations, from pledge to hostage, is perhaps indicative of how much had changed since the arrival of Le Sacre and La Pensée just a few days before. To be a pledge is to have an ambiguous status betwixt and between one who has been given and one who is possessed. A pledge is, quite literally, an oath enacted by a body – a living, breathing contract that can potentially bind togcther and ensure the good behavior, if not the interests, of the different sides represented in a transaction or encounter. The hostage, on the other hand, is

---

[106] Ibid., 43.
[107] Drakard, *A Kingdom of Words*, 49.
[108] Ibid., 196.
[109] Crignon, *Oeuvres*, 43.

an implied threat of loss. In both cases, positive or negative, promise or constraint, the path to surety – to confidence, or lack thereof – was built into a balancing act: hostages and pledges were to be held by both sides and be of commensurate weight. It was this balance that could maintain relations across cultural divides.

Parmentier's power, however, like that of the Chabandaire's, was limited by the position he held at the end of one of the scale's arms. In this sense, metrological agreement, closure, could not take place without the actions of intermediaries who had the power to shift the balance. Here "weights" asserted themselves into the negotiations. Once in place, they could assure, promise, explain, cajole, laugh, beg, and testify as to the reliability and trustworthiness of one side or another, and thus diffuse potentially dangerous situations to reestablish, if only temporarily, a balance of force.[110] In other words, it was words and promises, trust built up face to face through exchanged nervous laughter, the shared knowledge of immanent catastrophe, and ritually chewed betel, that brought them back from the brink of war; words thus supplemented the mensuration of exchanged bodies to help build trust across seemingly incommensurable divides. These instruments of credit, these exchanged bodies, could speak, question, cajole, intervene, lie, and persuade; in so doing they performed transvaluations of confidence where the authority of Parmentier acting on his own could not.

Confidence was thus a performance; it was the work of "binding" contracts and maritime insurance, but is was also the work done by astrolabes and chip logs, by well-armed and ordered men, and by pledges and hostages who were spies, ambassadors and balancing weights of surety. This distribution enabled them to navigate their ship across unknown seas, just as it provided a degree of surety in the precarious world of cross-cultural contact. But of course, it could also introduce new and dangerous contingencies, as the acts of the Scotsman, Vassé and Breant proved on Madagascar's southwest coast. Weights were not simply passive participants in the work of scale, they too could speak and act.

Violence had thus far been avoided, but now, without either pledges, hostages, or other agreed-upon weights and measures, on what basis could they continue their interaction?

---

[110] One could imagine, of course, other scenarios where pledges and hostages might exacerbate the tensions that occur when cultures collide; this would depend on individual interests and agendas that the hostages might harbor and pursue, or it might depend on their level of comprehension of – and ability to communicate with – those who were holding them hostage. In this particular case, it seems as if Crignon and the Portuguese were truly mediators – not just ambassadors for Parmentier, but agents in the possession of invaluable intelligence that could counsel their captain as to the best course of action on the basis of hard-won "expertise" in Minangkabau culture.

The next day, the Parmentier brothers returned to shore to ask the Chabandaire for the six gold marks they claimed that he owed them; he refused to hand them over. The captains were furious ("for other – unspecified – things as well"). They declared war on Ticou and demanded that the Chabandaire come before them. When he did not appear, the Parmentiers went to those they thought were in charge: the village Oranchaies, its leading citizens.[111] We do not learn anything about what transpired at this meeting, however. One can only surmise by Crignon's silence that the Oranchaies simply called the Parmentiers' bluff and that the French responded by cutting their losses and leaving. But this is only speculation, as Crignon's text suddenly diverts us from the consequences of Parmentiers' declaration of war, to a description of an encounter – which occurred at some unspecified point during their stay – that emphasized the incommensurable differences between the Ticouans and the French. Crignon thus shifts his narrative from recounting their difficult relations with the Ticouan harbor master and Parmentier's declaration of war to describing how Parmentier came to interrogate the high priest of Ticou about his knowledge of Christianity, and in particular, the complexities of the doctrine of the Trinity.

It is difficult to imagine the scene. Parmentier, accompanied by Crignon and his translator, Nicolas Bout, sought out Ticou's high priest to ask if he knew that "God had sent his divine word (*son Verbe*) making himself flesh on earth as incarnated by a Virgin through the operations of the Holy Spirit, and how this *Verbe*, who is his son, is engendered by the Father, while also being inscribed on the heart and mind (the soul, *la pensée*) of man?"[112] What the high priest of Ticou thought about all of this we do not know; indeed, he never even heard the question, for Nicolas Bout, Parmentier's *truchement* (his go-between translator), had absolutely no idea how to say this in Malay (*le truchement ne pouvoit bien parler de ces choses*).[113]

Jane Drakard describes the west Sumatran Minangkabau in the seventeenth century as living in "a kingdom of words." Words for them, she argues, instantiated the power of distant highland kings and their representatives. There was no sense that textual accounts or word-of-mouth reports of a king's feats, possessions or power were, or ever could be, fictional, metaphoric, or figural. Words in these circumstances were, she avers, quite literally, power; they were coextensive with and constitutive of reality – having the same status as Christ's real presence in

---

[111] Ibid.
[112] Crignon, *Oeuvres*, 44.
[113] Ibid.

the consecrated host for medieval Europeans.[114] Perhaps, then, had the archpriest and Parmentier not been limited to a practical, mostly ostensive, trade pidgin, they would have had much to say to one another.

Or would they?

The semiology of presence in medieval and early modern Europe was of course materialized in things, e.g. in relics, the Eucharist, and money, but it was also embodied in words, specialized ritual formula, oaths, signatures, seals and signs inscribed on instruments and books and, of course, on hearts and minds. It seems both the French and their Ticouan interlocutors lived in "kingdoms of words" as Drakard says, but this is where the similarity ends, for these kingdoms were inhabited by very different entanglements of humans, words, and things. To be sure, Drakard uses the presumption of similarity to explain a later difference – that between the "episteme" of the Minangkabau (which, she argues, maps closely onto the Medieval European theology of presence) and baffled post-Reformation Dutch Calvinists. It would appear, then, at least for the Dutch, that they weren't just encountering the Minangkabau on Sumatra, but their own medieval, pre-reformation, past! This being said, the impoverished view of the European past presented here ignores the long and complex history of resistance to – and critique of – sacramental notions of presence in the Eucharist and those who claimed to be its representatives. This is significant for a number of reasons. Succinctly, these critiques provoked increasingly articulate and evermore refined enunciations of orthodoxy, which in turn led to the targeting of critics and critiques as dangerous and heretical. Despite the best efforts of the Church, ambiguities associated with notions of literal presence and those who were its spokespersons continued to grow, while at the same time, finding both wider dissemination through the medium of print, as well as larger and more receptive audiences, especially in places like Normandy, which was known in Parmentier's day as a kind of "little Germany" because of the number of Lutherans said to be found there.[115] This is not to say that Parmentier was a Lutheran – indeed, in his poetry he compared Luther to a "vile sea monster" – but he didn't have to be, to be part of a world, and a milieu, where doubts about Christ's real presence in the Eucharist were finding increasingly clear and vocal articulation. Indeed, this was part of a much more widespread and far-reaching destabilization of – or rather, shift in – the theology of presence that had social, political and epistemic consequences far beyond simply

[114] Jane Drakard, *A Kingdom of Words*, 195–200.

[115] M. C. Oursel, "Notes pour servir à l'histoire de la Réforme en Normandie au temps de François Ier, principalement dans la diocèse de Rouen," *Mémoires de l'Académie Nationale des Sciences, Arts et Belle-Lettres de Caen* (1912): 139, 151.

putting into question the fidelity of Eucharistic transubstantiation. In this sense one can perhaps discern in the questioning of the doctrine of real presence, a more systemic representational shift toward understanding nature itself as the instantiation of God's presence in the world. Parmentier, for example, in his *Oration* can be seen to make this case in ways remarkably similar to that of his contemporaries Paracelsus and Cornelius Agrippa. The dissolution of the sacred and its migration from the sacramental into nature itself engendered, as a matter of course, new ways of speaking with, about, and for, God. The problem resided in the efficacy of this language and the social standing of those who were to speak it, which is to say, the social status of the new men and their knowledge of the contingent natural world had to be translated into terms consonant with the universal. Parmentier, knowingly or not, participated in this process – a process that disentangled the host's metonymic power from the validation of priests, prayers and alters and shifted it onto the new men, their instruments and nature. Increasingly, the relations between Heaven and earth were mediated not only by the words and gestures of a specialized bureaucratic cadre empowered to care for, "activate," and represent the sacraments,[116] but by astrolabes, compasses and maps; by well-ordered and knowing bodies (e.g. of astronomers, sailors, and artisans), and by the possession of specialized and recondite skills (e.g. of cosmographs, mathematicians, physicians, poets, and pilots). Not only did this work of metonymic displacement lend impetus to a stocktaking of the terrestrial world, it extended to similar efforts to understand and discipline vernacular linguistic conventions by which this world was to be known and described.[117] In other words, words themselves were treated in Parmentier's world as things that, like nature, could be distanciated on the page (or on brass), and standardized in systems of reference, not only in works of practical mathematics, anatomy, geography, and natural history, but also in books of customs, dress, and courtesy, and in works of lexicography, grammar, and rhetoric. This was paralleled by their material and institutional organization in cabinets of curiosity, libraries, archives, and centers of administration, such as Ango's headquarter in Dieppe, La Pensée, or the trésor of the l'église Saint-Jacques.[118] Words and things,

---

[116] See Wintroub, "Taking a Bow."

[117] See Wintroub, *A Savage Mirror*, esp. 167–192.

[118] See Michel Foucault, "Politics and the Study of Discourse," in Graham Burchell, Colin Gordon, and Peter Miller (eds.), *The Foucault Effect: Studies in Governmentality* (Chicago, 1991), 53–72; M. T. Clanchy, *From Memory to Written Record: England, 1066–1307*, 2nd edn (Oxford, 1993), esp. 156–164, and idem., "Does Writing Construct the State?" *Journal of Historical Sociology* 15 (2002): 68–70; Randolph Head, "Knowing Like a State: The Transformation of Political Knowledge in Swiss Archives, 1470–1770," *Journal of Modern History* 75 (2003): 745–782; and Mialet, *Hawking Incorporated*,

meaning and signification, were in this sense linked systemically through the collection and taxonomy of vernacular conventions that were disseminated as "trustworthy" standards in the same way as metrologies of weight and measure. This was necessarily an unending, fraught, and highly contested process. This reorientation of presence and the dissolution of God's body into nature itself, as well as the concomitant translations of social authority and prestige of those who could speak and write authoritatively about it, did not mean that Eucharistic ritual, Christ's real presence in the consecrated host, or priestly power, simply disappeared or faded away; indeed, the threat posed to the efficacy of the Eucharist as a vehicle of presence led to a kind of doubling-down on its metonymic power, hence the hardening of religious divisions from the 1530s. At the same time, the danger represented by these doubts served to add even greater urgency to attempts to establish – and represent – reliable and trustworthy standards. While the Minangkabau might have taken words very seriously, so too did the French. Indeed, Parmentier and his men regarded acts of mensuration with deadly seriousness.

The failure to find closure in the metrological realm was therefore closely intertwined with the failure to communicate across linguistic and cultural divides. Without the balance of hostages, let alone weights and measures, there was no reason or imperative to trust at all. Crignon thus proceeds – without transition and without returning to the surely delicate negotiations that had to be undertaken to avoid the war his captain had declared – to describe the Ticouans' homes, their physical characteristics and their clothing, before turning his attention to various aspects of the lives they lived, including the food they ate and how they slept. He then moves from this general description of everyday life among the Ticouans to his estimation of their character: they live lives more austere than men and women of the strictest religious orders back in France; "they aren't strong," he says, "but they are refined and clever (*astucieux*), great flatterers, great liars and deceivers, sometimes very amusing, always demanding. Had we done what they wanted," he explains, "we wouldn't have had any merchandise left to our names. Even after a deal had been struck, they would want to renegotiate for better terms or simply not honor their word at all. They would drive even the most level-headed (*sage*) man to distraction and fury." They were, moreover, "all like this," he says, "it was the custom of this land (*c'estoit la coutume du pays*), with the king and the greatest among them, all made from the same mold."[119]

---

chapter 6. On libraries, see for example P. Nelles, "Renaissance Libraries," Stam (ed.), *International Dictionary of Library History* (2001), 134–151; on cabinets of curiosity and their relationship to dictionaries, grammars and rhetorics, see Wintroub, "L'ordre du rituel et l'ordre des choses: l'entrée royale d'Henri II à Rouen (1550)," *Annales: Histoire, Sciences Sociales* 56 (2001): 479–505.

[119] Crignon, *Oeuvres*, 45.

Crignon thus takes us from a general description of the people of Ticou to a description of the actions of the man who was to become their syn-ecdoche, the Chabandaire. It was the Chabandaire who thwarted their every attempt at commerce; it was he who had poisoned their relations with all the merchants in the vicinity by forbidding any and all, upon pain of death, from buying French merchandise until he had made the first offer (which was always at "the worst possible price"). And this, Crignon explains, was why the captains hated him so much.[120]

The incommensurability of their encounter could not be overcome – indeed, those who spoke for kings and gods were not given to compro-mise. For a time Crignon and Antoine the Portuguese were able to ame-liorate the tensions and maintain a kind of precarious balance between the Parmentiers and the Chabandaire, but neither side was willing nor able to take advantage of this fragile truce to build a more lasting equi-librium. Each approached the interaction from the bluster of presumed superiority. One wanted to play an interminable game of back-and-forth haggling; the other wanted to set hard-and-fast terms linked to clearly demarcated material points of reference. Without the security of hostages held by both sides, they were incapable of even talking, let alone estab-lishing commercial relations.

Tipping the Scales

A sense of what might have occurred between the French and the Ticouans in the hours and days missing from Crignon's account (immediately following Parmentier's declaration of war) can perhaps be gleaned from a letter reproduced in the manuscript of Crignon's journal that Charles Schefer transcribed and published.[121] Dated June 18, 1575, it was from a Rouennais merchant named Guillaume Lefèvre; it recounts details related by an 80-year-old sailor named Jean Plastrier who participated in the Sumatran expedition. Plastier tells the story of hostages held on Le Sacre. It is relatively easy to fix the time of his account to the moment just before the French left Ticou. In other words, immediately following Parmentier's declaration of war and his visit with the Oranchaies.[122] It also suggests the likely outcome of this visit: namely, that the Oranchaies resisted Parmentier's demands and sided with the Chabandaire, and that Parmentier responded by taking

---

[120] Ibid., 46.
[121] Charles Schefer, *Le discours de la navigation de Jean et Raoul Parmentier de Dieppe: voyage à Sumatra en 1529: description de l'isle de Sainct-Dominigo* (Paris, 1883), 1–4.
[122] While at sea Crignon was very careful to date his log entries; after reaching Ticou dates are only occasionally indicated. The first hostage crises seems to have taken place some-time after November 14; the second, shortly before November 27 when they left.

hostages. Plastrier's account skips over all these details, and focuses instead on the fate of the Ticouan hostages now held by Parmentier. He describes their increasing agitation as they came to realize that the French were preparing to set sail. When night descended a few of them managed to escape on one of Le Sacre's large boats. This was considered a grave loss. Parmentier ordered the remaining hostages onto the other boat and gave chase. On approaching the shore he caught up with the escapees and made signs that if they didn't return he would execute the hostages he still held. When the now free hostages made clear that they would not return, Parmentier ordered that the heads of his prisoners be cut off; the master of Le Sacre dutifully complied. This, Plastier concludes, marked the definitive rupture in all relations between the French and the people of Ticou.

Parmentier's attempt to coerce the Ticouans clearly demonstrates that trust had been broken; surety, insofar as there could be any, needed to be compelled through leverage – by bodies held hostage. "I have your people, and if you don't do as I say, you won't see them alive again." The balance being weighed here was not an attempt to mediate between the interests of Minangkabau traders and French merchants, but between Parmentier's need for control and Minangkabau resistance. Taking hostages would, in theory, ensure his and his crews' safety, while also giving him the leverage he needed to win compliance for his demands. Perhaps the most interesting element of what appears to be a quite straightforward asymmetric assertion of power is the way it operated through a series of rhetorical reversals to shift the balance of responsibility for the lives and deaths of the hostages away from those holding the guns and the knives to the people they were kidnapped from. In other words, in the moral economy of hostage taking, it was up to the Oranchaies to decide what was going to happen to their people, not Parmentier, who was, by this logic, a neutral and objective instrument calibrated to the Oranchaies's (or the escaped hostages') "good" behavior. But weights, as we have seen, have the power to assert themselves, and hostages don't always act like hostages. Jugurtha is a good example of this (see Chapter 3); so too, of course, were the hostages that escaped from Le Sacre. Had Parmentier wanted to maintain his power over the Minangkabau he would have surely killed the remaining hostages one at a time. This, at least, would have demonstrated his "seriousness" while leaving the door open to further negotiations. The murder of all the hostages at once, however, signals his abandonment of the search for either balance or leverage. Parmentier had become unbalanced. Perhaps like the Malagasy with the bloody shirt, his ruthlessness speaks to his desire to demonstrate his power through an act of collective punishment – they *were* the people of

Ticou that he hated so much for his loss of face, and of course, for the loss of one of his boats.

The murder of the remaining hostages points to Parmentier's frustration, his reactionary rage and his violent assertion of (his very limited) power. Indeed, violence frequently supplemented and underscored Parmentier's authority, whether by implication, as with his triumphal arrival in Ticou, with the killing of hostages on Le Sacre, or the post-mortem dissections of the son of Pontillon and the Breton sailor, Jean Dresaulx. But while balance might had been abandoned with regard to the Minangkabau, it was very much an ongoing concern on board Le Sacre and La Pensée, with his men facing the ruin of their high hopes and the dashing of their expectation that they would return to Dieppe with holds filled with gold and spice. The killing of the hostages was not a message directed at Ticou, since that encounter was already dead; it was, rather, a message directed at Parmentier's own crew. In other words, killing his hostages was a compensatory act of surety meant to counter, through the demonstrative display of vicious authority, the failures of his leadership and the desultory end to which he brought them. The blood of hostages was, perhaps, the only "pixie dust" still left to him.[123]

---

[123] One wonders if Crignon found this either convincing or laudatory, hence the omission in his Log of any mention of this incident.

6   Replication: Replicating a Thought

Replication is to make the same thing in another place. Put somewhat differently, replication aims to duplicate the conditions that frame the creation of an original – an event, an experiment, a play, an object – so as to do it again, and again, and again. This meaning, however, along with its resonance with scientific notions of truth, came relatively late in the word's history, in the early twentieth century, though the sense that it would then acquire was clearly in the air in the seventeenth, describing well the replication of experimental findings in the 1660s by the Royal Society's program for the production and legitimation of matters of fact by repeated trials.[1]

Replication has now become a commonly accepted criterion for the determination of truth in science. According to Karl Popper,

only when certain events recur in accordance with rules or regularities, as is the case with repeatable experiments, can our observations be tested – in principle – by anyone. We do not take even our own observations quite seriously, or accept them as scientific observations, until we have repeated and tested them.[2]

In other words, replication works both to create stability and truth – proof – and to standardize human understanding and practice such that one human can, in theory, stand (as a witness) for any other if the appropriate recipes are followed and conditions met. This interchange-ability – this inter-subjective consensus – is built, says Popper, from repetition, from "regularity and reproducibility." As he puts this: "Only by such repetitions can we convince ourselves that we are not dealing with a mere isolated 'coincidence', but with events which, on account of their regularity and reproducibility, are in principle inter-subjectively test-able."[3] However, as Harry Collins has shown, it is problematic to assume that there could ever be an account or a recipe of some original event or action made in such detail, and with such precision, that it could

[1] See Shapin and Schaffer, *Leviathan*.
[2] Karl Popper, *The Logic of Scientific Discovery* (London and New York, 1959), 23.
[3] Ibid.

be followed and exactly replicated elsewhere. Indeed, the empirical study of replication – as a social and material process – shows it to be extraordinarily complex, tacit, improvisational and inexact, whatever its putative epistemic (namely social) leverage for notions of scientific truth.[4] This lack of determinacy takes us back to earlier meanings of the word *replicacion* or *réplique*: a refrain, an echo, a reflection of an original, a copy, and at the same time a carrying over (a translation) of an original into new domains of resistance and improvisation – another experience, another voyage, another stanza, but more of a transformation than an iteration, more a reply, a response or a retort to an original, than a replicated copy.[5] After all, one does not respond with a mimicking echo to a question or a statement; that is, by simply repeating the same words back again. This would be nonsense. Responses, like replications, are made with reference to something said or done – they are formulated out of tangled interests and desires, and from acts of appropriation, adaption, transformation, negotiation, and resistance.[6]

Though the meaning of *réplique* typically had to do with the act of replying, especially in disagreement or defense, or in a legal context with a reply to a plea or charge, it also had more concrete references: to turn, fold, bend, unroll, or unwind, and by extension, the right to claim restitution, recovery, or restoration. As in the tension between its almost contradictory meanings to fold or unroll, to turn or unwind, we once again find ourselves on a kind of seesaw – a balance tipping this way and that. On the one hand, restoration is an eruption of the past into the present to recover or restore, a labored erasure of the process of time to establish the return of an idea, a thing, or a process to an "original" state or condition; on the other, it is an expedition of the present into the past to set things right by demanding *restitution* for actions – or wrongs – done. Replication, in all these senses, is about readjusting the balance between the past and the present, selves and others, here and there, one side and the other.

Replication – *Replicare, Replico* – is composed of the roots *Re* + *plicare*. Re: about, regarding or with reference to, and plicare: to fold, entwine, entangle. Re, as it is commonly used, indicates a displacement, a shift

---

[4] See Harry Collins, *Changing Order: Replication and Induction in Scientific Practice* (Chicago, 1985); see also Simon Schaffer, "Glass Works: Newton's Prisms and the Uses of Experiment" in D. Gooding, T. Pinch and S. Schaffer (eds.), *The Uses of Experiment* (Cambridge, 1989), 67–104; O. Sibum, "Reworking the Mechanical Value of Heat: Instruments of Precision and Gestures of Accuracy in Early Victorian England" in *Studies in History and Philosophy of Science* 26 (1995): 73–106.

[5] See ibid., esp. Collins.

[6] In this context, M. Bakhtin is worth consulting: *The Dialogic Imagination: Four Essays*. Trans. and ed. M. Holquist (Austin and London, 1981).

that takes us to another time, place or medium – "this" is about "that," where "that" could be literally anything – words, people, things, a situation, a context, a metaphor, etc. Re thus signifies a relationship or an identity – a continuation of (or a relation to) something in another place. In some cases, this reference – this referral – is a citation, a grounding in proof, a testament of authenticity by reference to validating evidence that is found elsewhere. In its plural form Re is *Res*: wealth, goods, concrete things and substantive matter. This matter, of course, was also the stuff of ideas – that is, of word-things as understood, for example, by Erasmus's literary enactment of Quintilian's notion of the "treasure house of things" in his *Collectanea Adagiorum* (1500) or his *De Utraque Verborum ac Rerum Copia* (1512), or in Joachim du Bellay's *La Deffence et Illustration de la Langue Francoyse* (1549).[7]

In his translation of Sallust's *Catiline Conspiracy*, for example, Parmentier argues that "it is not enough to read [ancient authors such as Thucydides, Herodotus, Sallust and Livy] one must attentively *revolvée, calculée et relevés* [them] to know their profundity and to understand their substance."[8] As the trinity defining his mode of study implies, ancient texts were for him material objects that were to be picked up, handled, turned around, and examined from all sides (*revolvée*); measured, gauged, deliberated upon, reckoned and calculated (*calculée*); and copied, responded to, and redeployed (*relevé*). There was much of the merchant about Parmentier's relations with the ancients; but these relations also fall squarely in the humanist tradition exemplified by the likes of Erasmus and Du Bellay in viewing classical texts as the matter and the substance – the Res – of invention.

Relevé is particularly interesting; among its many meanings, Cotgrave includes that of a Saint "whose Reliques, or bones are inclosed and shewed aloft, in a shrine."[9] But it also encompassed a range of other important meanings and associations: to make landfall and go ashore; to collect, take down, note, or copy; to recover, or to establish a proof in a difficult situation; to react, answer or respond;[10] and, by the seventeenth century, the determination of the position of a cape or an island and the act of inscribing this on a map.[11] Relevé thus tracks closely with and recapitulates many of the manifold associations of "replication," similarly attempting to put a name to the complex and often contradictory

[7] See Terence Cave, *The Cornucopia Text*, 19–21; and Wintroub, "The Looking Glass."
[8] Parmentier, *Oeuvres*,130, from a note to his readers, in Parmentier's *Traduction du Catilina de Salluste*.
[9] Cotgrave, *A Dictionarie*, s.v. *relevé*.
[10] See the *Trésor de la langue française*, s.v. *relever*.
[11] Ibid.

interactions that subsist among acts of collecting, copying, reference (reverence), transformation and resistance (treason) to words, texts and peoples at a distance (whether, temporal, social or spatial).

Words are enfolded into things and vice versa.[12] Rabelais taught us this on the Frozen Sea with the thawing sounds of the battle fought only months before between the Arimaspians and the Nephelibates.[13] The pairing of *re* with *plicare* thus seems only natural: to fold, entwine, envelop embrace, to clasp, associate, or implicate. All these meanings touch on the folds of a supple cloth as it bends back upon itself to repeat – or metaphorically refer to – in manifold iterations and multiple copies (*plicare*). Separately, and together, moreover, *re* (*res*) + *plicare* refer to authenticity – to proof, as for example in a reference or citation, or in the fold of a document demarcating the site where a seal is to be affixed, or to where a validating signature is to be placed.

As with scale, there is a sense that discipline is implicit in such replicatory acts; indeed, "replication" or "replicare" was closely related to the Middle French *repeticion* (repetition) – the act of stating something again and again, of going back or returning to, a request for repayment, or a claim of the right to have something back, as in the prayers repeated by monks across Europe measuring the day with books of hours and rosaries beads as they followed – observed – their daily routine: Lauds, Matins, Prime, Sext, Nones, Terce, Vespers, Compline. Repetition was (and still is) a rehearsal, a practiced iteration enacted to fix a ritual, a play, a series of words or gestures in the memory of bodies and minds. Replicare or replico has many cognates; among the most relevant in the present context are *explico* (to unfold, unfurl uncoil, to disentangle, to solve, settle, arrange, and regulate); *implico* (to entangle, entwine, enfold, embrace, associate, and implicate); *multiplico* (to multiply and extend); and *supplico* (to pray, supplicate, beg, plead, or entreat).

Regulate, implicate, extend, and entreat – this describes well the workings of replication. The essence of prayer, for example, is repetition: a ritual form of communication which builds its efficacy from the material of disciplined minds (and bodies) that can produce repeated iterations of a series of words to construct not only a persuasive interaction with God (*commercium*), but an induction (an entanglement implicating and initiating) the praying supplicant into a community of likeminded believers. Here words would be calibrated to rituals, gestures, and things – to

---

[12] Not only with regard to the materiality of printed words, but to the materiality of words as they are spoken and understood – that is, embodied, as living acting things generative of, and generated by, the language games within which they are enunciated, understood, and transformed.

[13] See the Introduction.

repetitions followed in a text, or counted out and logged by prayer beads (*patenostres*).[14] These material measures of "induction" (in the sense of becoming part of) were not only markers of prayers, but of the calibration of individuals into networked communities held together by common notions of regularity, repetition, and stability. Perhaps, then, patenostres featured so prominently as items of trade on Parmentier's expedition for reasons that go beyond mere aesthetics, their low cost, and the ease with which they could be stowed and transported.

Beyond formal acts of repetition, replication thus referred to memories fixed and plotted on minds and extended in bodies, and to "intersubjective" processes of thinking and judgment, that is, "replique" – to ponder, think about or reflect upon. But as we have seen before, this was not simply a matter of copying – of duplication. Mental replicas – of an external world, whether as perception, memory, or rule – were "inexact" and "imperfect," and had to be constantly checked, negotiated, and calibrated against the beliefs and actions of others. These intersubjective negotiations were materially coordinated, amplified, and, of course, locally inflected, by purposefully made and constructed things. The making or shaping – *facere* – of things (as opposed to finding them), draws us back to replication in the sense of multiplication and repetition of "artificial" things molded, formed or created by human action such that one could stand in for the other as credible copies. Replication, in this sense, had two senses, the replication of things, and the replication of the dispositions (explicit and tacit, cognitive, social, and moral) that were entangled in them. As with any replication, these multiplications were contested, negotiated, resisted, and inflected. The frontiers – between man-made and natural objects, the authentic and the inauthentic, the true and the artificial – mapped out by facere and its cognates *factice* (factitious) and *fabricāre* (*forgier, forger*) create a continuum stretching from produced, constructed, and composed, to forged, fabricated, and fictional. Was a replica true, a corruption, an idol, a counterfeit? Indeed, how could the "same" thing *be* in another place or time?

Amongst the earliest mass manufactured objects that replicated – copied – one another were coins and the Eucharist, both of which were made in increasing numbers from the thirteenth century. Coins, like hosts, were meant to be exact replicas of one another, with each individual instance capable, in theory, of substituting for any other from the same batch. Made from the same mold and according to the same template,

---

[14] Prayer beads were first mentioned in the eleventh century and commonly used from the twelfth. John D. Miller, *Beads and Prayers: The Rosary in History and Devotion* (London, 2002).

they were meant to be identical. One can imagine how much work had to be done to ensure that these made things could be used as stand-ins for the presence God and sovereign authority. Part of this work was to be found in the folding and enlisting (replicating) – by factions, interests, institutions – of supplicants and citizens into the formation (and extension) of larger communities that shared understandings and expectations about them. However, translation is always accompanied by those who traduce – that is, by treason. The work of establishing precise mechanisms by which authenticity was to be adjudicated in the field was arduous to say the least. The money of account was in theory all the same, but money in circulation lived in a much more precarious and uncertain world; coins were counterfeited, clipped, rubbed, and washed. The same held for hosts, which were desecrated, challenged, misused, and misunderstood. Just as coins were regulated by officials of the mint, masters, moneyers, assayers, and clerks of the market, hosts were only hosts if made, cared for and used by authorized delegates of the Church. But how were these agents of authentication themselves to be trained and policed? How could one ensure the legitimacy of – and compliance to – their authority as spokespersons? Fidelity of replication was not a given, it had to be ensured through constant, careful, and often violent, expressions of discipline.[15]

If money and hosts were difficult to replicate in the more easily controlled conditions of monasteries, churches and mints, imagine how difficult their circulation in the wild must have been. Voyages were indeed difficult to replicate. In the case of the circulation of ships, specific – if flawed – recipes found in logs, maps and navigational treatises were important resources in regimes of replication; these were supplemented by the tacit knowledge of seamen whose experience could be exploited, modified, transformed, taught, and transferred. However, even if this technical and embodied knowledge could be reliably communicated in a statistical sense as "good enough," this would not have been enough to constitute replication. Replication, as the word's original usage makes clear, is a response, a retort, or a reply. Replication thus refers to a relation, a generative transformation, that picks up and reacts to – adjusts, grows out of, and interacts with – prior statements, contingent and always changing natural conditions, previous voyages, and earlier experiences (namely, experiments). Parmentier's *Oration* perhaps exemplifies a dimension of this, as he sought to fold seagoing knowledge, erudite humanist sensibilities, recondite mathematical skill, and religious devotion into an expression of his creditable

[15] See Wintroub, "Translations."

leadership in rhyme. Sailors needed to be constantly checked, policed, and calibrated through shipboard rituals, interactions with others, with things, with nature, and with demonstrative acts of leadership, whether with words, lashes, postmortem dissections, or the death of hostages. By means of the constraints provided by belief, habit, and fear, tired, hungry, and self-serving sailors could be transformed into a crew – a cohesive unit, an escale, a host, capable of sustaining a voyage across the world. In other words, crews needed to be inducted, recruited, con-scripted, and engaged. This complex and dynamic coordination of men, ideas, practices and things, was the condition of replication, not its con-sequence.[16] Rather than growing out of replication, as is often thought, consensus could only grow with it. Perhaps this is why it is so difficult to achieve and maintain. Discipline, motivation, incentive, habit, and inspi-ration thus form a crucial part of any recipe of induction – *implico* – the enfolding of individuals into an escale capable of replication.

For perhaps obvious reasons, we should linger on "induction" for a moment. Estienne defined "induction" as persuasion.[17] Cotgrave defined it both as "an inducement, allurement, or persuasion, unto" and also, "a forme of argument from particulars to universals."[18] The solidification of the meaning of induction as a form of reasoning based on the gathering of evidence in support of a larger claim seems to have taken place around the middle of the sixteenth century. The rhetorical and the epistemo-logical thus form two sides of the same coin, each playing a crucial role in the process of replication. But between these Janus-faced meanings, stands an array of others: thus, Estienne cites Cicero in defining *inductio* as a carrying of water from one place to another, and to the translation of the soul, thought, or heart (*l'esprit*). The journey from the particular to the universal is thus accompanied by real and metaphorical voyages, a fluid carrying over of the supplicant into the concord of an escale, a host, a team, a crew. Estienne, however, also defined inductio as an efface-ment or canceling out of a text (*escriture*), presumably by this flowing of water.[19] This meaning finds an internal referent in the word's use – in the effacement (and black boxing) of the work of translation. Thus rhe-torical aspects of persuasion, recruitment, and inauguration were elided and cut off from the idea of induction as proof based on the matter of particulars, on res. This canceling out is of course nominal not "real," as is amply evinced by recent work by historians of science and scholars

---

[16] On the experimenter's regress, H.M. Collins, "'Son of Seven Sexes', The Social Destruction of a Physical Phenomenon," *Social Studies of Science* 11 (1981): 33–62.

[17] Estienne, *Dictionarium*, s.v. *inductio*.

[18] Cotgrave, *A Dictionarie*, s.v. *induction*.

[19] Estienne, *Dictionarium*, s.v. *inductio*.

of STS. Certainty (even in a probable sense), has as much to do with inclusion, recruitment and initiation, as with res. As Thomas Wilson aptly put this (c. 1560): "we mighte heape many men together, and prove by large rehersall, any thyng that we would, the whiche of the logicians is called induction."[20] In other words, once again, we find that consensus does not result from putative acts of replication (of rehearsal, *répétition*, *réplicasion*), but vice versa.

We thus begin to see emerge – across a history of related words, *replicare, facere, inductio, exemplum* – a discourse of particulars made into universals that were able to persuade and inaugurate through the erasure (*inductio*) of a range of associated meanings having to do with forgery, counterfeiting, and rhetorical persuasion. Replication is here being born in a modern sense. What is particularly interesting is how processes of effacement work to extract and exclude from the meanings and practices associated with these and related words all traces of response, reply, retort, persuasion, feigning, and fiction – thus effectively hiding the fallibility, interests, and contingent ambiguities that were necessarily part of their fashioning. Particulars could, accordingly, be persuasively arranged as the material of a new kind of universal – i.e. facts, specimens, instances, and cases – to form the matter (*res*) upon which compliance, confidence, and consensus could be forged.[21]

The steps in La Pensée's and Le Sacre's journey previously outlined – the intelligence gathered and mobilized before they left (Chapter 1); the socio-technical expertise displayed and conjoined to forms of popular piety (Chapters 2 and 3); the disciplined enforcement (namely, objectification) of social order enacted by on-board anatomy theaters (Chapter 3); the scaling of abstract knowledge and its transfer to mobile bodies and media (Chapter 4); and the attempts to work out and discipline metrologies of navigation, contact, and leadership (Chapters 5 and 6) – recapitulate processes of replication: the calibration and coordination of different voices (interests) so that they could speak and act (reply, retort, riposte) together as one. This enfolding – this "implication" – of ship, crew, navigators, and captains into a force capable of enacting a balance against wind, seas, stars, and foreign peoples, required more than logs, maps, astrolabes, and nautical skill, it also required, as we saw in Chapter 1, inspiration, motivation, and desire. This was true not only for the voyage itself, but for the voyages that were to follow. One needed reasons to go, and to go again; these ranged from those

---

[20] Cited in the OED, s.v. induction.
[21] Surely this is what Latour is pointing to when he argues that we have never been modern.

given by Parmentier in his *Oration* – i.e. for God, king and country – to those exemplified by his men's reckless pursuit of gold and ginger on Madagascar.

For status conscious "new men" like Parmentier and Crignon, commerce was as desired as it was troubling; on the one hand, it defined their status, power, and authority; on the other, there was something menial, lower-class, unsavory, even ungodly, about it. In other words, commerce for them had to be translated into the language of the educated, cultured, pious, and devout in the same way that contingent particulars had to be reframed in terms of universal truths rather than sublunary happenstance. Going across the world in search of gold and spice was risky and uncertain business. In his poetry, and in his onboard sermon in verse, Parmentier struggled to translate the desire for profit and glory into the language of a spiritual cause, a quest. Replication was, after all, repetition; it was a rehearsal for the next try, for further trials, for other assays. It was also a working out of reasons for why to go again. Voyages like Parmentier's were meant to be inaugural – to begin a process of emulation and repetition. They spawned recapitulations (*récapituler*) such as Crignon's Log, his eulogies to Parmentier, the *Discours d'un Grand Captaine de Dieppe* published by Ramusio in 1539, and maps by Fine and Rotz, etc. A copy, it should be recalled, was an *exemplum*: something taken out and brought forward – a sample, a case, a specimen, a fact, a text, an incident, a person, or even a voyage, that exemplified, or evinced qualities to be learned and standards to be imitated.[22] As we will see in the pages that follow, the life of Parmentier's voyage – translated, multiplied, and circulated in print, on maps, and on instruments – was just such an exemplum.

La Pensée's and Le Sacre's voyage of return from Sumatra back to Dieppe will begin this chapter; tributes to Parmentier will stand at its middle, and the reframing and migration of the voyage into travel accounts, maps, and new forms – and disciplines – of knowledge will compose its end. In this regard, it is perhaps relevant to note that just as "scale" in the sixteenth century referred to ports-of-call on a journey, "replication" also referred to the practices of mariners. Thus, in addition to its other meanings, "plicare" meant both the packing of a ship's sails so that they would be ready for the next voyage, and to a ship's return from sea to land.

---

[22] This conforms closely to Kuhn's understanding of the word; see Thomas Kuhn, *The Essential Tension: Selected Studies in Scientific Tradition and Change* (Chicago and London, 1977), xix. See also, John Lyons, *Exemplum* (Princeton, 1989).

The Scales of Death

They left on November 27. Many were sick and dying, and many had died already. The first to begin "the dance" and cross over (*trespasser*) was Jean Parmentier. He passed on November 3, on the eve of Sainte Barbe, eight days after becoming sick. It is difficult to chart Parmentier's illness with regard to his actions in executing his hostages. Dates scrupulously tracked at sea in Crignon's Log disappeared once they made landfall on All Saint's Day. Time did not seem to matter in the same way on solid ground as it did at sea where the days, stars, winds, and tides were constantly, and carefully, monitored and calibrated with instruments, ephemerides, calendars, and charts. It is impossible to say precisely when Parmentier fell ill, but given his behavior, one could speculate that he had already become symptomatic (*fievre, chaud mal ou fleux*) when he so unceremoniously executed his Minagkabau hostages. With many of the crew sick and close to death, they followed the coast south, sending the small boats to look for supplies when they could. They were told by some of the natives that there was pepper to be found in Indapoure (Inderapura) about a quarter day's sailing south. On November 23 Raoul Parmentier died; that same day, Jean Masson went to shore to negotiate for pepper. Despite all that had befallen them, they maintained their devotion to commerce, and to returning to Dieppe with the spice that animated their journey.

Though the French "treated them kindly" (*amoureusement*), the natives ran away each time they approached. Crignon speculated that lies had been told about them. Recall, however, that he would entirely expunge any mention of Parmentier's treatment of the Ticouan hostages from his account of the voyage. Unfounded rumors and lies were the only explanation left to him. Of course, it is far more likely that word of the treacherous Frenchmen had spread.

Seeing that there was no pepper to be had, they had to decide what to do. Some wanted to continue to Java, others wanted to return to Indapoure or to Priame to sell their wares and find pepper. Others wanted to give up altogether and go home. It was therefore decided to get the opinion (*avis*) of the entire *communauté* – that is, to discuss and poll – the men of both ships to decide what to do next.

Trust and confidence, as we have seen, were distributed across multiple bodies and then given a name – a name endowed with specific attributes of authority and expertise – in this case, La Pensée's captain and the expedition's leader, Jean Parmentier.[23] Power was built into the

---

[23] This tracks closely to Mialet's work on leadership, genius and the distributed-centered subject; see Mialet, *L'entreprise*, and *Hawking Incorporated*.

coordination of many to act, and speak, as one.[24] This distribution was constructed not only across the competencies, skills, and expertise of the crew, but across generations of shipboard tradition and hegemonic ideologies of social hierarchy; it was also built out of necessity. Indeed, without a certain degree of concord and coordination, ships could never successfully sail over such long distances. However, reliance on land-based social order would only take them so far; indeed, authority on board ships was far less reliant on birth or networks of patronage than on the performance of deeds (and, of course, the ability to spin them appropriately). In other words, authority was contingent on the success-ful choreography – and packaging – of a crew's skill in sailing a ship from one point on earth to another. Failure would result in death – either by the seas or by mutiny. Daily life and habitual routine aided in this per-formance, maintaining the system's coordination on a kind of autopilot of shipboard rituals and traditions. Without consensus, the fragile bal-ance between the ship and the forces of nature could not be maintained. There were no possibilities of retreat or escape. Intra-crew strife would have disastrous consequences. As Michel Serres has noted, collectivity was "enclosed by the strict definition of the guardrails."[25] Nevertheless, active intervention – such as Parmentier's *Oration*, his postmortem examinations, his insistence upon the reliability of his own geographical knowledge over that of his Portuguese "expert," his decision not to pur-sue (as his men wanted) the promise of gold in Madagascar, or his execu-tion of his Minangkabau hostages – could be required at any moment.[26] Shipboard life was not simply a question of knowing when and how to sail, it also entailed deciding where to sail and why. Leadership was a skill that was relatively autonomous from the crew's skill in navigating a ship; which is to say, a good leader was not necessarily a good sailor, and vice versa. However, the respect of the crew required a certain flu-ency in the language and the bodily skills of sailing; the coordination of this with more esoteric forms of knowledge might well impress, if not compel, compliance. Though the crew might be able to navigate with-out a leader, the stability of such a ship would be precarious to say the least. Leadership was a crucial element of the assemblage of practices and things that worked to restrain, channel, coordinate, and motivate

---

[24] Ibid.

[25] Michel Serres, *The Natural Contract* (Ann Arbor, 1992), 40.

[26] Judging by other evidence, discipline was often maintained with the most draconian methods. That Crignon does not mention them in his Log does not mean that they did not exist. As we have seen, he was not shy about leaving out material that could tarnish Parmentier's image. See, for example, Cheryl A. Fury, *Tides in the Affairs of Men: The Social History of Elizabethan Seamen, 1580–1603* (Westport, CT, 2001), 46.

a crew. It was not necessarily a necessary condition, but depending on the conditions, sometimes it was. In a crisis, the captain's actions could serve to shore up 1) his own leadership (his postmortem exams); 2) the motivation of his crew (his *Oration*); 3) his men's ability to balance out the threat posed by natives (forming them into a well-ordered and triumphant host); or 4) his crew's resistance to the lure of profit when its pursuit would be too risky (Madagascar). Even captains had to work hard to convince and implicate (induct) crews into making, and investors in investing, in long and uncertain voyages. What then would happen if the body lost its head? What would become of La Pensée when it was no longer being led by Parmentier's "authoritative" brain?

This simple question raises a host of others: could the cooperative distribution and coordination of competencies among the crew be preserved without someone in charge? Were longstanding and ingrained rituals of cooperative interaction able to maintain themselves independent of a fixed social hierarchy and clearly established leaders? To what extent was the sailing of a ship and the successful coordination of its crew an attributive – or claimed – possession of its captain, rather than a cause? How much was autopilot – experience and tacit knowledge of ships, oceans, winds, and shipmates? How much was hands-on skill in leading? Was the captain, to ask the question in a slightly different way, the only member of the ship's company capable of seeing the crew's actions at a distance – as a manipulable abstraction, a kind of interactive workflow mapping that only he could access by virtue of the social (and cognitive) distance that his position commanded? Or was this something that was implicit – tacit – in the work itself and independent of its theoretical and social articulations? If the latter was the case, could the system be maintained over time without the generative social fiction of leadership – without a head to speak for the body?

Consultations

In the sixteenth century crews were often consulted with regard to major decisions.[27] The pursuit of harmony – of concord – was of crucial importance not only for the success of a voyage, but also for survival.[28] For this reason, it was not uncommon for crews to be lied to, hoodwinked, and deceived.[29] This being said, the life-and-death stakes, the ingrained skills, the habits of cooperation developed over months working together

---

[27] Ibid., esp. chapter 2, and idem "The Work of G. V. Scammell," in C. Fury (ed.), *The Social History of English Seamen, 1485–1649* (Woodbridge, UK, 2012), 27–46, at 43.
[28] Ibid., 46.
[29] The case of Columbus' relations with his crew come to mind.

closely, could help overcome the pursuit of individual interests, and the flaring of petty enmities and resentments. Following traditions of military organization, ships were, at least ideally, self-regulating systems of integrated and coordinated action (*une escale*) in balancing out the threatening powers of nature and "Others." We can, in this sense, discern in the webs of mutual dependency spun in navigating ships across long distances a kind of proto-democratic deliberative process – what Serres refers to as a "natural contract." At the same time, one wonders, in the face of evidence supplied by Plastrier, if this view is too sanguine. As he put it, having been left with no master, everything fell into a state of disorder (*le tout fut en désordre*); so it was decided to return as quickly as possible to France.[30] The crews of La Pensée and Le Sacre did not, however, entirely disentangle into the conflicting pursuit of self-interest with Parmentier's death, as a cursory reading of Plastrier might lead one to believe; indeed, Plastrier soon qualifies his earlier statement, writing that the highest-ranking members of the crew then "took the advice of the two ships" (*prirent avis aux dits deux navires*), and "it was decided" (*ce qui fut acordé*) "to return to France."[31] Crignon's account provides more details, stressing not disorder, but deliberation, strong opinions, dispiritedness, and delegation.

According to the Log, on November 28 Guillaume Sapin, La Pensée's contremaistre, Jean Le Roux and Crignon went to see, hear, and "enregistrer" the deliberations on board Le Sacre. Pierre Mauclerc and Antoine de La Sarde, the maistres of the Sacre made beautiful speeches (*remonstrances*) to the men that enumerated the trials and tribulations they had suffered on their journey: they had lost their captains, two contremaistres, many good companions, and their large boat; many had died, and many were sick and in danger of dying; their supplies were all but exhausted, the ships were leaking, and the weather threatened to delay their departure for another seven or eight months. After the speeches were made, "thirteen or fourteen" of the men voted to return immediately to France, while "nine or ten" wanted to persevere and make for Java. That same day, after dinner, Mauclerc and de La Sarde came to La Pensée; but only "two or three" of the men wanted to discuss where to go; all they really cared about was finding food and fresh water. They thus decided to lift anchor and set sail back toward Indapoure to look for supplies and to try to salvage their voyage with one final attempt to trade with the natives for gold and pepper.

---

[30] Schefer, *Le discours*, 3.
[31] Ibid.

A Last Chance

Jean Masson went to Indapoure's port on November 29 with a sample selection of their wares; he was told there was no pepper to be had, but that they could trade for gold. They bartered; agreements were made, and orders were taken. The next day they brought their merchandise and were told that no trade could take place without the leave of their king and a gift for the Chabandaire. The French complained bitterly, arguing that they had settled on a price and had delivered the merchandise as they said they would, and that they had been "honest and forthright" in all their dealings with them, and that they had no cause to complain. In the end, their arguments must have proved persuasive, as they sold 8 measures of fabric for a "facet," and a bougran for two "coupons."[32] On the first day of the New Year they sent Le Sacre's remaining boat to shore to look for food, trying to preserve what little they had by eating only rice and water. They stayed until November 18, trading "rouge de Paris, cloth, bourgrans, mirrors and paternosters for gold, rice, honey, poultry and 2 behars of pepper."[33] Several more men died while they were there, and for this, among other (unnamed) reasons, they set sail, on January 22, for France, thus bringing Crignon's account to a close.

A Poet's Lament and the Narrative Redemption of a Journey

They arrived in Dieppe some time in May or June. Plastrier provides a few further details about their return journey. They stopped around the Cape of Good Hope for about a month, and at Saint Hélène's, an island off the west coast of Africa. They "rescued" six "Indians" marooned there by the Portuguese. One of these castaway "savages,"

---

[32] Schefer uses *facel*, and assumes that this is a mistake in transcription or orthography and should read tael. The tael, however, was a measure of weight usually associated with spice or metal, rather than cloth. A coupan is a small copper coin of variable value. Schefer, *Le discours*, 3.

[33] Ibid. Schefer, quoting M. l'abbé Favre's Malay-French dictionary defines bahar as the "nom d'un poids qui varie selon les lieux et les choses que l'on pèse. Dans beaucoup d'endroits, le bahar de clous de girofle est de 550 livres, celui de muscadest de 57 livres, tandis que celui de poivre serait seulement de trois pikul ou 375 livres." *Dictionnaire malais-français*, Vol. 1 (Paris, 1875), 181, au mot Bahara. According to Vincent Leblanc, again quoted by Scheffer, with regards to pepper "Le Bahar...est de trois cent soixante livres, peut valoir trois escus et demy ou quatre au plus fort: ce qui peut revenir à un Lucatan ou cinquante-cinq sols le quintal." See, *Les voyages fameux du Sieur V. Leblanc* (1648), 138.

Plastier notes, married a French women and died in 1569, almost 40 years later.[34]

On January 7, 1531, Crignon published, with Gérard Morrhy, the *Description nouvelle des merveilles de ce monde et de la dignité de l'homme*. Appearing not more than eight months after his return, this little book was an explicit attempt to salvage the voyage after the fact by reformulating the reasons it was undertaken and rehabilitating the Parmentiers' sullied reputations. As such, it was the first, tentative, step in the process of replicating the voyage of La Pensée and Le Sacre. As Crignon said in his prologue:

> as one of his closest and most intimate friends, for the delight of all noble and virtuous souls enamored in seeing and hearing word of the world's description (*cosmographie*) in this to contemplate and heed the marvels that God has made in heaven, on earth and at sea; and wanting to submit to the urgent (*importune*) requests of friends and family to reduce (*rediger*) to writing his navigation and voyage, and to describe and bring it to light so that the name of the Parmentiers would not be buried with their bodies on the island of Sumatra, but that they triumph over death and be brought back into the memory of men by their renown and immortal glory.[35]

The book consisted of a collection of Parmentier's verse, including the *Oration* and several of his prize-winning poems, such as his "*Au parfaict port de salut et de joye*," for which he won the "crown" at Dieppe's Puy de l'Assomption in 1527. Crignon clearly wanted to enfold himself into the reputation he was attempting to build for his captain, adding his own works to those of Parmentier, such as his elegiac poem, the *Déploration sur la mort desditz Parmentiers*, which he wrote shortly after Raoul Parmentier's death while they were in Indapoure.

Crignon's lament begins with a problem. The voyage was an utter and complete failure that did not, as he put it, "result in any praiseworthy consequences (*louable effaict*)." Gold and spice eluded them. Had the Parmentier brothers lived, he says, things would have been different. However, without them, and with the men hungry, sick, and dying, the ships' new master (its *patron*) decided to make a hasty return without engaging in any further exploration or trade.[36] While Crignon's Log intimated that decisions were made collectively in consultation with the crews of both ships, his poetic eulogy indicates that the Parmentiers' leadership had been replaced

---

[34] Ibid., Thevet recounts a similar story; see Jean-Claude Laborie, Frank Lestringant (eds.), *Histoire d'André Thevet Angoumoisin, cosmographe du roy, de deux voyages par luy faits aux Indes australes, et occidentales* (Geneva, 2006), 119–120.

[35] Parmentier, *Oeuvres*, 2–3.

[36] Crignon, *Oeuvres*, 55.

by an unnamed "patron" (perhaps Mauclerc) who was not at all shy about expressing his disdain for the expedition's former leader: "deriding, indeed, hating all that the deceased captain loved; not being able to stand (even) the sight of Parmentier's poor dog, nor even myself, though I did everything I could to dutifully carry out his orders."[37] It seems that directing the crew's frustrations at the now dead captain was an effective means of consolidating the new patron's leadership. Much of Crignon's poetic eulogy was dedicated toward rehabilitating Parmentier's reputation in the face of this resentment. He begins by attempting to humanize Parmentier and his brother by recalling the sad scene on Dieppe's docks when they took leave of their wives (their "nymphs"). Thus he describes Jean's wife,

> Whom he loved with all his heart
> Saying a farewell (*adieu*) so filled with sorrow and regret
> That a heart harder than hard
> Would split (*fendu*), or melt (*fondu*) like wax to see it
> Never was a goodbye so difficult to say.

In placing himself as a witness to her pleas, Crignon uses her tearful goodbye as a vehicle to express his own misgivings about their arduous undertaking, providing a very different answer to the question asked and answered by Parmentier in his ship board *Oration*: "What am I doing here?"

"How could you want to leave your 2 children (*enfans laisser*)," she says, "for a country so far away with which to converse (*converser*)?" Perhaps referring to the lost expeditions of Pierre Caunay and Jean de Breuilly sent to the Indies by Ango in 1526 and 1528 respectively, Parmentier's wife pointedly asked her husband if anyone had ever returned from such a voyage. The question was rhetorical, everyone in Dieppe surely knew that no one ever had. She then directly addressed her husband's reason for undertaking his voyage to the Indes. "Desire for honor makes you do this," she said, foreshadowing "Reason's" intervention in Parmentier's *Oration*. Recall Reason's words:

> For sake of honor
> As the first Frenchman to undertake
> A voyage to a land so faraway
> ………………

---

[37] Ibid., 56–57. Dogs were often kept as mascots on ships; for example, the skeletal remains of a dog were found amongst those of the human crew of the Mary Rose, sunk in 1546 in Portsmouth; see http://news.bbc.co.uk/2/hi/uk_news/england/hampshire/8564209. stm. More interesting, perhaps, given the hatred evoked by Parmentier's dog, is the close association of dogs with scholars and the scholarly habitus – the companion canine forming part of an image cultivated by intellectuals, and constituting part of the standard repertoire furnishing the humanist studio.

> Your enterprise is for the glory of the king
> For the honor of your country and for yourself.[38]

But unlike "Reason" in Parmentier's *Oration*, his wife was not an ethereal muse calmly advising her husband from another world. The tearful context implies a far different tone – one of reproach and resignation. She continues in this vein:

> Would to God that your so lofty heart,
> Would do my bidding: it would then know the truth.
> Do we not have goods enough
> To live together in joy and happiness
> Without giving you either pain or care?
> ....
> If it would please you to follow my will
> You would not have to go to the Indes
> Or expose yourself to such great danger
> To acquire things of such little value.

Thus, she sadly chastises her husband, challenging his assertion in the opening lines of his *Oration* that he did not undertake his voyage across the world for riches or self-aggrandizing honors, but for God, king and country.

Crignon, the author of this poetic history, expresses a similar view in a chant royal he wrote not long after his return to Dieppe, which he included in his elegiac collection of his friend's verse.

> Once when humans lived in great pleasure
> In a place of peace and delectation,
> Neither cold nor heat could cause them harm.
> But falling amorous of every good,
> Proud desire of ardent affection
> Placed in their hearts through its subtle language [the desire]
> To go to sea to take up navigation.
> In the orient where precious gold is to be found
> And through its profitable commerce
> Loaded with all manner of alluring goods
> To the great profit of all the public good.
>
> Humans hearing this vain boasting
> And therefore navigation undertook

---

[38] Ibid., 92.

> 'Quand ce vouloir te esprit,
> De te donner tant curïeuse peine,
> Cela tu feis, afin que honneur te prit
> Comme françoys, qui, premier, entreprit
> De parvenir à terre si loingtaine;
> ...
> Tu l'entreprins à la gloire du roy,
> Pour faire honneur au pays et à toy.'

And put to sea bodies, goods, provisions and sustenance,
Raised anchor and without delay
They hoisted high lofty sails.
The wind of pride like thunder and lightning
Carried them off with such violence
That it drove their ship to sundry places
Far from God toward the Antarctic pole
Holding always with the wind better to find
To the great profit of all the public good.

And when they finally found land
Proud Atropos by mortal command
Showed them the results of his power,
Making their bodies putrefy
With signs of his infection
Even the strongest of this noble crew
From the noblest to the insignificant page.
Upon seeing this had been heedful
Of turning towards heaven's benevolence
This furious danger to avoid,
To the great profit of all the public good.

And despite waves of ignorance,
With jealous and obstinate winds,
Until the cape called good hope
They sailed to their salvation;
And while navigating by the elevation
Of the true sun, by whose shadow they steered,
So that they had come to a safe and beautiful place,
Where with a fair sky, clear and without mist,
They saw rise beneath the oblique horizon
The pole star with luminous rays,
To the great profit of all the public good.

This constellation gave assurance
To the navigators in distress
That they would soon arrive in the land of France
With joy and consolation
The sickly for comfort
Took the bread of life as a necessity
That gave them health and more;
They would drink the most gracious wine of love
Each man applied himself to praising God
And from the skies the Virgin descended
To the great profit of all the public good.

Prince we will honour the courageous Virgin,
Who preserved from deadly shipwrecks.
Navigators in perilous dangers;

> We will sing her magnificent virtues,
> For she has marvelously saved us
> To the great profit of all the public good.[39]

Echoing the words he gave to Parmentier's wife, Crignon reflects upon and reassesses his recent adventures in Sumatra as a means of recalibrating voyages of exploration and commerce as religious pilgrimages. Commerce, he argues, is part of a sinful world. The unanimity found in shared Marian devotion was both a retrospective desire of what should have happened on their voyage and an exemplum conducive to future navigational and entrepreneurial exploits. Commerce was thus to become *commercium*, which after its liturgical usage, was communication with God. This is perhaps why Crignon worked so hard to publish Parmentier's *Oration*, which similarly aimed to transform their expedition to the Indes into a spiritual quest.

However, the death of the Parmentiers could only serve as an exemplum of "the great profit of all the public good" if their reputation could be restored; that is, if Crignon could formulate an appropriate response (*réplique*) to rectify the wrongs he felt had been done to his friend's honor and reputation. Madame Parmentier thus reminds Jean (and Crignon, his readers) how fickle men could be, especially when success and the public good were judged solely in terms of acquisitive desire and worldly ambition:[40]

> If your labour is reversed by fortune,
> Adding to your work and pain,
> You will be hated and disdained
> By all who flatter and adore you,
> And as much as they had sung your praises
> They will come to quickly blame you;
> After the sweet you will taste the bitter
> ....
> If all you promise does not come to pass.[41]

We sense here, perhaps, how close La Pensée was to mutiny before Parmentier died. Using the Captain's poor widow to voice his own doubts, as well as his own pain at having lost his friend, Crignon draws the parallel between Parmentier parting from his wife and his own experience in saying goodbye to his dying companion. His verse pulls Parmentier down from his perch in the ship's hierarchy to make him into a living, breathing, suffering human being saying goodbye to a wife he would perhaps never see again, rather than the distant, erudite "brain" at the helm of La Pensée.

---

[39] Crignon, *Oeuvres*, 65–66.
[40] Note the resonance of these ideas with those articulated by Sallust regarding the decline of Rome.
[41] Crignon, *Oeuvres*, 57.

He didn't know how to comfort her

…

His heart broken, and his eyes brimming with tears,
He regretted his decision to undertake this voyage,
Not knowing what to say, his voice was taken
Between breaths parting his heart
Which was losing its strength and vigor
Then a kiss she gave him in the end
And a barely audible goodbye
Like the sound of a sorrowful shadow
Presaging the funeral
That was to come.[42]

Crignon's mind next wanders from this human world into the world of the gods, and from the past back to the present, closing his eyes on the ship sailing toward France, he falls to sleep, and is visited by Morpheus in the form of poetry. He dreams and Morpheus speaks:

Don't run from or resent
That which pleases the divine goodness;
Think always of accomplishing his will.
……………………………………
[Death] may have put an end to the Parmentier brothers' lofty
    (*haultaines*) works
But the good God who was their fiancé [their trust, credit, assurance,
    their confidence]
Does not want to let their good names perish in the flood of forgetting
Along with the death of their bodies.
It is His pleasure that a record be made [re + cord]
Of their many honorable virtues
And that praise be made for their souls to inherit
And be lifted into the heavens and transformed as stars
With the brightness
And the beauty of Castor and Pollux.[43]

Here things become complicated. Morpheus tells Crignon that he needs to commemorate the life and work of the Parmentier brothers by lifting the "polluting fog" that has obscured their brilliant light. This fog, he says, is like the paint that a "perfect painter" applies to a rotten piece of wood. The beautiful and life-like image, however, dissolves, when the paint is scraped off.

A great and perfect painter makes
A rotten, dirty and infected piece of wood,
Seem beautiful with a little paint
That he applies to it

[42] Ibid., 58.
[43] Ibid., 60.

To make an image
So lifelike that he will do homage to it
And we will all say: this is a living thing!
But if any dispute or quarrel arises,
He [the painter?] will know how to establish and prove
That beneath [the paint], it is a rotten piece of wood that he will find.[44]

The analogy is odd, and indeed, a rather backhanded way of saving the Parmentiers' reputation. How could a beautiful painting be construed as polluting fog, and how could a rotten piece of wood be compared to the Parmentiers' "splendid light?" Going out on a limb, one might speculate that Crignon is here referring to the Parmentiers' rotting corpses, to their bodies that had been interred on land and at sea, and by extension to the vanity of praising their worldly deeds. The painted wood thus was a kind of idol, a mask hiding the realities of the sinful material world. In this regard Crignon's funereal *Plaint* closely parallels the narratives carved into stone of *transi* tombs such as those made for the Fuggers in Augsburg, or – closer to home – that made by the atelier of Antonio et Giovanni di Giusto di Betti (Antoine et Jean Juste) for Louis XII and Anne de Bretagne at the Abbey of Saint Denis in 1531, the same year that Crignon published his poetic obituary for the Parmentiers.[45]

The design of the tomb of Louis XII has often been credited to Jean Perréal, someone who – as we have seen (Chapter 4) – was surely well known to Parmentier and Crignon through mutual connections to Rouen's Puy de palinod.[46] Crignon's literary monument to his dead companions shares much with these stone memorials. Transi tombs consisted of realistic figurations of the body immediately after death, or as a decayed cadaver. Louis XII and his wife Anne, for example, were depicted with lips drawn back and hollowed cheeks, skin carved taut, withered, and drawn tight against protruding bones and distended

---

[44] Ibid.
[45] See Anatole de Montaiglon, "La sculpture française à la Renaissance: La famille des Justes en France," *Gazette des Beaux-Arts* 12 (November, 1875); also see Pierre Pradel, *Michel Colombe, le dernier imagier gothique* (Paris, 1953); Erwin Panofsky, *Tomb Sculpture. Four Lectures on its Changing Aspects* (New York, 1964); Jean-Marie Jenn, Françoise Jenn, Jean-Pierre Babylon, and Alain Erlande-Bradenbourg (eds.), *Le roi, la sculpture et la mort: Gisants et tombeaux de la Basiliqe de Saint Denis*, in *Archives départmentales de la Seine-Saint-Denis, Bulletin* 3 (June, 1975); and Kathleen Cohen, *Metamorphosis of a Death Symbol: The Transi Tomb in the Late Middle Ages…* (Berkeley and Los Angeles, 1973), esp. 133–180.
[46] On Perréal's involvement with the design of Louis XII's *transi* tomb see Barbara Hochstetler Meyer, "Jean Perréal and Portraits of Louis XII," *The Journal of the Walters Art Gallery* 40 (1982): 41–56; M. de Laborde, *La renaissance des arts a la cour de France*, t. I (Paris, 1853), 186; and R. de Maulde la Clavière, *Jean Perréal dit Jean de Paris: peintre de Charles VIII, de Louis XII et de François I*[er] (Paris, 1898).

organs, all held together by the stitches left by the royal embalmers (see Figure 6.1). Such realistic representations of death were not simply *memento mori*; as their name implies, transi were depictions of a movement – a translation – between earth and heaven, from death to eternal life (or in the case of kings, from the mortal body to immortal *dignitas*), as represented by a second tier, built above the representation of the corpse, upon which idealized representations of the body when alive were depicted devoutly praying (see Figure 6.2).[47] Corruption was, in this sense, a necessary step in a journey toward spiritual regeneration. Glorious deeds, material success, pride, and honor – as represented, for example, in the famous bas reliefs depicting Louis' triumphs in Italy on his tomb – were harnessed to spiritual movement, propelling the souls responsible for them to the next, divine, level as depicted by their living figurations praying above. Such worldly accomplishment would be but vanity and wind – painted words – if they were not entangled in this religious narrative. Thus Crignon's *Plaint* criticizes the "flatterers, sycophants, poisonous detractors" among the crew who "painted a great lie of the truth (*paindre un grant mensonge de verité*)," by spreading vile rumors and invective against the dead, while "vainly praising" them to the skies while they were alive. There is a thing called truth – the rotted wood – that exists beneath the painted lies, but there was also a sacred rhetoric that could be employed, as the gods directed Crignon to do, to redeem the Parmentiers' worldly deeds by transforming them into a prayer to the Virgin and to God; thus would his friends' corrupted flesh be resurrected in a more pure and eternal form.

At the end of his poetic lament Crignon finds himself on an exotic island inhabited by a host of different gods. He is greeted by Polymnie (the Muse of Rhetoric). She recounts that she has known them all from a young age; that she has loved and nourished them with the "fluent milk of her white breasts," and that it "was his [Crignon's] responsibility, as a friend to Jean and Raoul, to describe with honest praise the rewards they will receive for their honor and virtue."[48] The gods thus transformed Jean's "rotting" corpse into a palm, which produced wine as sweet as the words he uttered while alive. The palm was, as we have seen, a ubiquitous symbol of triumph, closely allied not only with Roman rites of triumph and medieval and early modern royal entries, but with Christ's entry into Jerusalem.[49] They were utilized, moreover, by Normandy's Puys to

---

[47] See E. Kantorowicz, *The King's Two Bodies: A Study in Medieval Political Theology* (Princeton, 1957).
[48] Crignon, *Oeuvres*, 62.
[49] See Chapter 1.

Figure 6.1 Jean Juste, Detail of Louis XII's and Anne de Bretagne's effigies, lower tier of their transi tomb, Basilique Saint-Denis (completed 1531).

denote first place in their annual poetry competitions for the best chant royal; a prize redeemable for the not inconsiderable sum of 100 sols tournois (approximately 5 livres). Parmentier's transfiguration and apotheosis into the veritable sign of triumph can be seen as a fulfilment of the promise Crignon made in his preface to the *Description nouvelle des merveilles* that "the name of the Parmentiers would not be buried with their bodies on the island of Sumatra, but that they would triumph over death and be brought back to the memory of men by their immortal fame and glory." And indeed, Raoul, who had been buried at sea, was transformed into a similarly potent symbol, a dolphin signifying the Christian purity of the French royalty, here inflected toward the imperial extension of France's power into seas and lands claimed by the Portuguese and the Spanish. Thus, as the "king of fishes" with a white cross on his back, and carrying a lily, he was tasked with guiding the French on their voyages across a sea named to preserve forever his and his brother's memory: the sea of *Parmentiere*.[50]

[50] The proto-nationalist exemplum provided by the Parmentiers in Crignon's *Plaincte* was matched by their role as municipal boosters in the anonymous "*Autre plainte sur*

Figure 6.2 Jean Juste, Transi tomb of Louis XII and Anne de Bretagne, Basilique Saint-Denis.

la mort de Jean et de Raoul Parmentier," which was included in Crignon's small volume. This poetic eulogy compared the Parmentiers's inventive excellence in composing "ballades, chantz royaulx, moralitéz, comedies, rondeaulx, astrolabes, spheres et mapemonde, aussi cartes pour congnoistre le monde," to the works of Dedalus, Zeusis (Zeuxis) and Praxitelles, ancient artists all well known in the sixteenth century for their unparalleled mimetic abilities. Unlike their classical forbearers, however, the

The Death of a Poet

Despite this praise, and Crignon's efforts, the Parmentiers' reputation faded; but this did not mean that the work of replication stopped. Jean lived on, for a short while at least, as a respected poet.[51] However, changing styles and fashions soon judged the rhetoriqueur poetry he wrote to be labored, awkward, cumbersome, and hopelessly outdated. As Du Bellay put this in his *Defense and Illustration of the French Language*: "leave the old fashioned French poetry to the flowery games of Toulouse and the Puy de Rouen, such as rondeaux, ballades, virelais, chants royaux, chansons and other such spices (*espiceries*), which corrupt the taste of our language, and serve only to bear witness to our ignorance."[52] Yet, in a certain sense, Parmentier must have been an odd in-between figure for Du Bellay, for while he believed Parmentier's sort of "affected" poetry to be pernicious, he surely would have admired his wide-ranging maritime exploits. Indeed, Du Bellay's *Defense and Illustration* was modeled after a similar voyage of exploration, though one in time not in space. As he put it, "fearing that the winds of affection would push my ship so far out to sea that I would be in danger of being shipwrecked, I retook the route that I had original set" – a route directed toward ancient Greece and Rome in the hunt for models that could be creatively utilized to enrich and improve the French language.[53] Du Bellay's replication of Parmentier was thus, at one and the same time, a rejection and a reaction (*une réplique*) that propelled him away from the affected "spice" of the *rhetoriquers* toward new poetic and rhetorical forms; it did this, moreover, through a kind of emulation of

anonymous author argues that the Parmentiers put their talent to work in the service of the Virgin Mary – thus winning renown for Dieppe and upholding its reputation as second to none in "grace, biens et richesse." [Parmentier, *Oeuvres*, 142–143]. Ferrand attributes the "*Autre plainte sur la mort de Jean et de Raoul Parmentier,*" to Gérard Morrhy. This is unlikely. There is little reason to think that Morrhy had any investment in Dieppe, or that he would be so assertive in its promotion and defense. Morrhy's signed contribution to Crignon's collection, "*Epitaphium Joannis Parmenteirii, qui in Samothracia periit,*" immediately following this "*autre plainte,*" is much less personal and lacks reference to Dieppe. The folio pages that contain his verses, moreover, are in a different type font than the rest of the book, perhaps indicating that his contribution was added quickly, perhaps as an afterthought, or late arrival. Indeed, the reference to Samothracia, which is a Greek island in the northern Aegean, in its title perhaps indicates how peripheral he was to Crignon and the Parmentiers, and how easily memory could be transformed.

51 See Émile Picot (ed.), *Théatre mystique de Pierre du Val et des libertins spirituels de Rouen, au XVIᵉ siècle* (Paris, 1882), 89–90.
52 Du Bellay, *La defense*, 125.
53 Ibid., 136, ff.

Parmentier's voyages of discovery, not to new worlds across the seas, but across the frontiers of time.

The Mapping of Silence

We see similar processes of forgetting, elision, and replicatory transformation in the domains of cartography and maritime exploration, where Parmentier's reputation as a mariner suffered much the same fate as his reputation as a poet. In Oronce Fine's *Nova, et integra universi orbis*, for example, Sumatra is almost entirely obscured from view by the not-so-subtle placing of heraldic devices associated with the Valois dynasty (the emblem of Francis I, the salamander, holding up the shield of his first son François, the dauphin, surmounted by an imperial crown) directly at the island's middle with its extremities neatly divided across the two halves of the map. Sumatra is marked by its erasure, while at the same time providing the site for a symbolic claim to ownership. Thus, though Parmentier disappears, his encounter from the year before survives in Fine's demarcation and elision of Sumatra, its name (Taprobana) clearly visible in spite of its bisection, as a French (a Valois) possession (see Figures 6.3 and 6.4).[54]

Similarly, though Parmentier is not specifically mentioned anywhere in Rotz's *Boke of Idrography* (c. 1534), given to Henry VIII in 1542, his voyage can clearly be tracked across its charts. The first stop made by Le Sacre and La Pensée was the unnamed island across from Fuegos (the island of fire, and Rotz prime meridian); this appears on Rotz's chart as Saint Jacques, and is known on Portuguese maps as Santiago (see Figure 6.5).

Their next escale, Saint Laurent, or Madagscar, explicitly references the voyage, depicting in harrowing terms the fatal French-Malagasy encounter of July 28, 1529 (see Figure 6.6).

Following the Sacre and the Pensée travelling northeast now, we next find them near the "archipelague de Calicut" (the "archipelago of Callicut" on Rotz's chart), at "0 degrees under the line" according to Crignon's Log. On September 19 they were near the archipelague d'aupres Calicut et Commori – the Maldives; on the 20th they were ½ degree south of the line, landing on "Moluque." On Rotz's chart the Maldives appear, as claimed by Parmentier, well north of the line, not below (as Antoine the Portuguese maintained). The name of Moluque

---

[54] As the poet cosmographer Jean Maillard said a few years later, "*France feust maintenant a ses isles / Ou portugays ont place primeraine….*" BN Ms fr. 1382, *Description de tous les portz de mer de l'univers*, fol. 3ᵛ, my emphasis.

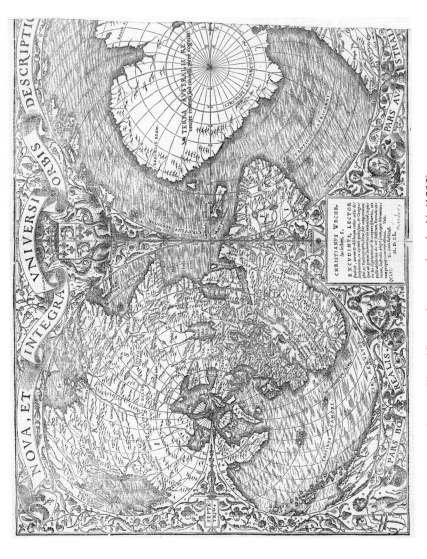

Figure 6.3 Oronce Fine, *Nova, et integra universi orbis* (1532).

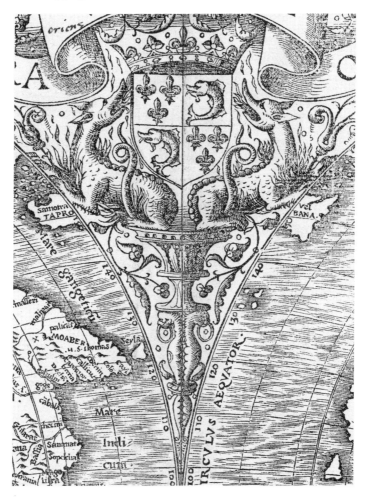

Figure 6.4 Detail, Oronce Fine, *Nova, et integra universi orbis* (1532).

is corrected, however, avoiding any confusion with les Moluques, as
Pouwa Moloku. The position of this island on Rotz's chart conforms
closely to that found by Crignon, about half a degree south of the line
(note that Rotz's charts are oriented with the South at the top) (see
Figure 6.7).

The voyage continued south, south-east, hugging the latitudes immediately below the equator, weaving north through a series of small islands,
which they named Louise, Parmentiere and Marguerite. They anchored,

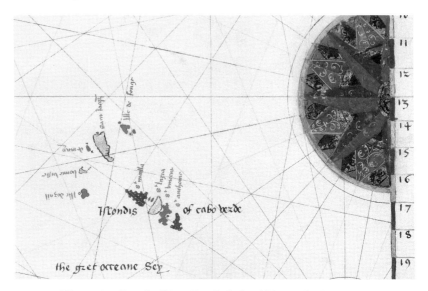

Figure 6.5 Detail of Jean Rotz's *Boke of Idrography* (c. 1534).

two leagues from land. Ticou is depicted on the west coast precisely where the east-north-east wind rose bisects the island. Interestingly, the Sumatrans portrayed on Rotz's Sumatra have their backs turned to the west, to face (greet or challenge) the east (the Strait of Malacca). Perhaps this was a commentary on the cold reception the French received in Ticou (see Figure 6.8).

On their return journey, they stopped at the Cape of Good Hope, seeing "a great troop of horned beasts like buffalo [or] cows and a number of men who watched over them"(see Figure 6.9). Recall that Plastrier's account appears in the historical record only in 1575, while the maps in Rotz's *Boke* were produced in the 1530s. Rotz's maps relied on other sources; in his treatise on magnetic variation (1542) he claims that he had spoken with, and obtained information from, "several of those who had made the voyage to Trapobane."[55] Thus we find a pictorial representation on Rotz's chart that closely anticipates (and perhaps provides a model for) Plastrier.[56]

---

[55] As paraphrased by Helen Wallis, "several of those who made the voyage from Dieppe to Taprobane, he writes, "have reported that they found a variation of 35 degrees east at Taprobane. But they had sailed so far from the diametrical line (the line of zero variation, which Rotz held to go through the island of Fuegos) that he did not know if there was an increase to a higher figure and then a decrease to 35 in the course of the 125 degrees of longitude sailed." See Wallis, *The Maps*, 6.

[56] For Plastrier, see Crignon, *Oeuvres*, 48–49, or Schefer, *Le discours*, 4.

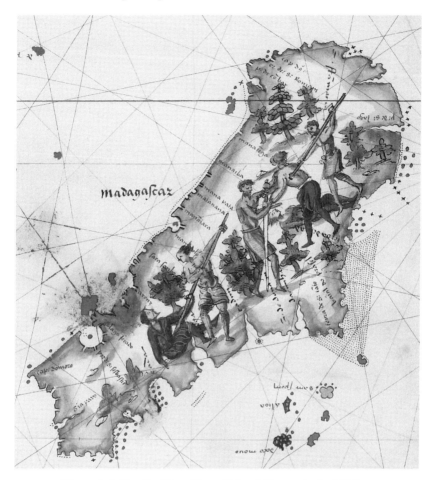

Figure 6.6 Detail of Jean Rotz, *Boke of Idrography* (c. 1534).

A Great Captain Speaks

Contemporaneous with the production of Rotz's *Boke of idrography*, an anonymous manuscript was written that recounted the expeditions of a "great captain" from Dieppe to "terre neuve des Indes occidentales" (or "New France"), and to Brazil, Guinea, Africa, and the Indes orientales. The original is lost, but the text was published by Ramusio in 1556.[57]

---

[57] G. B. Ramusio, *Navigationi et Viaggi*, volume 3 (Venice, 1556).

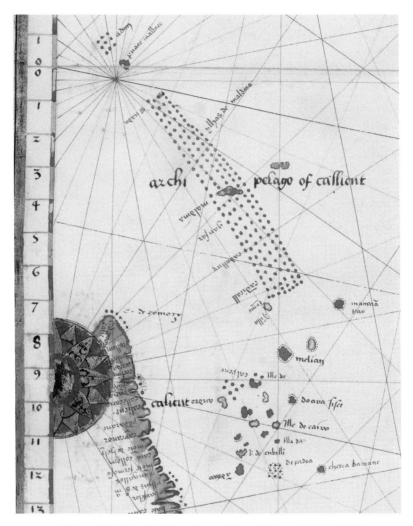

Figure 6.7 Detail of Jean Rotz, *Boke of Idrography* (c. 1534).

This "great captain," Ramusio tells his readers, led "2 ships armed in Dieppe to Taprobane in the east, today called Sumatra." There he "negotiated with the natives, and loaded his vessel with spice before returning home." "This discourse," he continues, "is truly very beautiful and worthy of being read by all." The work is unsigned and the original has been lost; it was not definitively associated with Parmentier's voyage until

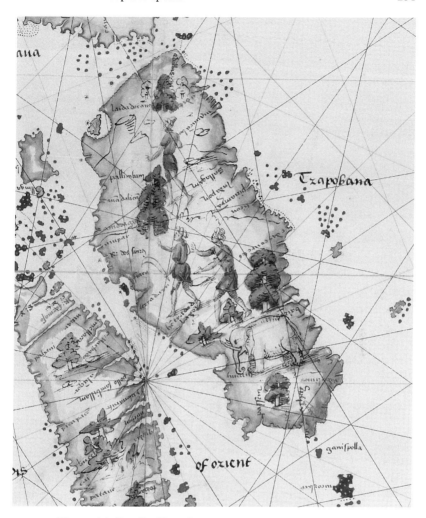

Figure 6.8 Detail of Jean Rotz, *Boke of Idrography* (c. 1534).

the nineteenth century, when Crignon's Log was rediscovered. Since then, the *Discourse of the Great Captain* has been attributed to Crignon on the basis of "word-for-word" correspondences with Crignon's Log. Upon closer examination, however, it appears that the two texts differ in fundamental ways, especially with regard to the very different accounts they provide of the encounter between the French and the Minangkabau.

Figure 6.9 Detail of Jean Rotz, *Boke of Idrography* (c. 1534).

According to the Great Captain, the people of Ticou "are very good
and peaceful people, but very clever and shrewd in trade and in business." Moreover, he says, "they keep their word in all agreements." He
continues:

I personally have had no dealing with them with the exception of two officials...
One of these ... was a captain of the army called Nacanda Raia, which means
'Captain of the King.' The other, who was called Cambendare, set the prices on

all merchandise that we brought there and distributed it to the merchants of the country. He made quick and good payments to us. No one dared to buy before the said Cambendare set the prices under penalty of decapitation. ... The cambendare [also] collected the king's money and tribute that was levied on each product sold. This tribute amounted to 3 per cent of the cost.[58]

The Great Captain of Dieppe ends his account with a description of Sumatran weights and measures, and by painting an idealized picture of his expedition's success. "After loading our ships with peuere (pepper) and other merchandise (*speciere*), we returned to Dieppe after a long and dangerous voyage made to honor God and the Crown of France."[59]

Compare this account of a voyage to Sumatra with the difficult, and ultimately, fatal encounter described in Crignon's Log and Plastrier's letter. Recall how much Parmentier hated and mistrusted the Chabandaire, and Crignon's less than flattering portrait of the Ticouans as absolutely unreliable and untrustworthy. The Chabandaire and Parmentier agreed about little, and most certainly not about weights and measures, or the percentage of tribute to be paid. Parmentier's voyage, as recounted in Crignon's Log, ended in disarray and failure – with the complete breakdown of relations between the French and the Sumatrans, a declaration of war, and a hold all but empty of cargo. In the account by the Great Captain from Dieppe, on the other hand, the voyage was an unqualified success, marked by a profitable voyage to Sumatra, good relations across huge cultural differences, a hold full of pepper and other merchandise, a dangerous but safe return to Dieppe, detailed intelligence about the land, its peoples and their commodities, and of course, the glorification of God and king.

Another, much more successful, voyage to Sumatra – to Ticou – in the years immediately after the return of Le Sacre and La Pensée, would provide a ready explanation for the Great Captain's upbeat account of his encounter with the Sumatrans. While not impossible, this is unlikely. A secret on this scale would have been difficult to hide, and indeed there is no documentation or reference anywhere of such a voyage, either directly, by rumor or by allusion.[60] Perhaps Crignon was writing about the brief period after the death of the Parmentiers when the French

---

[58] Hoffman, 28.
[59] I rely on Ramusio's version rather than Hoffman's translation here; see Hoffman, 40.
[60] Demarquets speaks of an expedition mounted in 1531 to follow up on Parmentier's voyage to China (sic), where it traded for porcelain, tea and other goods. This was written in 1575 and, as Roncière has pointed out, there remains no trace of such a voyage; see *Histoire*, 3: 277.

traded in Indapoure. Here too, however, the Log's description is very different from that of the Great Captain. Crignon's Log thus recounts difficult negotiations with the natives, and relatively meager trading for "gold, rice, honey, chickens and coqs, and 2 behars of pepper."[61] Did returning to Dieppe with 2 bahars of pepper constitute a successful voyage? Hardly. According to L'Abbé Favre's *Dictionnaire malaise-français*, written in the last quarter of the nineteenth century, "a bahar ... of pepper is only 3 pikul or 375 pounds."[62] Two centuries before, a bahar of pepper weighed in at 360 pounds,[63] and a century before that, it was, according to Tomé Pirés (writing in c. 1512), about three quintals and thirty arrates (approximately 330 pounds).[64] What was the measure of success when speaking of a cargo of pepper?[65] The Vicompte de L'Eau of Rouen registered 19 balles and 60 pounds (about 3,860 pounds) of pepper arriving in 1477–8.[66] In 1544, André de Malvende, a Rouennais merchant, received 56 balles (approximately 11,200 pounds) of spice, pepper, and ginger.[67] Plastrier claims that Le Sacre and La Pensée loaded up with some 30 tons before returning to Dieppe. A return after such a long and arduous journey with less than 700 pounds (2 behars) of pepper would hardly have constituted a success either to be bragged about or emulated.[68] The replication of the voyage thus alters and transforms the narrative of Le Sacre's and La Pensée's voyage from 1530 to 1539, converting its failure into an unambiguous success, while simultaneously effacing the presence of Parmentier, his brother, the taking and executing of hostages, and the meager cargo with which they returned. If the text was by Crignon (and indeed, if it was about Parmentier's journey at all), the poetic promise to commemorate his life and deeds was forgotten in favor of an account that eschewed authorship, origins, and emblematic leaders, in favor of a disembodied and distanced view of the Sumatrans and the good relations the French had with them.

[61] Crignon, *Oeuvres*, 48.

[62] Quoted in Schefer, *Le discours*, 86; see note 33 of the present chapter.

[63] Quoted in Schefer, V. Leblanc, *Les voyages fameux du Sieur V. Leblanc* (1648), 138.

[64] Tomé Pires, *The Suma Oriental of Tome Pires. Account of the East, from the Red Sea to China, written in Malacca and India in 1512–1515*, Armando Cortesao (ed.), 2 volumes (New Delhi, 2005), I:86.

[65] According to Lach, relying on evidence supplied by Luzzato, "in the decade after 1505, Portuguese vessels annually brought into Lisbon spice cargoes averaging from 25,000 to 30,000 hundredweight." Lach, *Asia in the Making*, 119.

[66] Mollat, *Commerce Maritime*, 230.

[67] Edouard Gosselin and Charles de Beaurepaire, *Documents authentiques*, 86. For a slightly later period, see H. Kellenbenz, "Autour de 1600: le commerce du poivre des Fugger et le marché international du poivre," *Annales. Économies, Sociétés, Civilisations* 11:1 (1956): 1–28.

[68] Schefer, *Le discours*, 3.

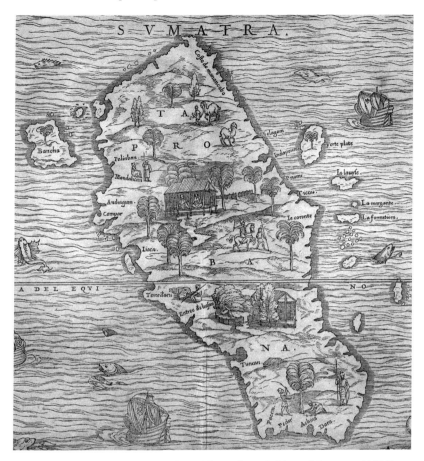

Figure 6.10 Giacomo Gastaldi, attributed, Map of Sumatra (1556).

Did Crignon write this very different account of his voyage to Sumatra? Possibly. One clue is the map by Giacomo Gastaldi included with the Great Captain's text (see Figure 6.10). This shows familiarity either with Crignon's Log (which was not published until the nineteenth century), or possibly a map by his hand, as the names that were given to the three islands off Sumatra's western coast were – though orthographically confused – similar to those given in Crignon's Log: La Lauyse, La Marguerite, and La Formetiera.[69] The Ramusio map is the only (surviving) cartographic representation of these islands so named.

---

[69] This is generally thought to be an engraver's error, for La Parmentiere.

Perhaps the disparities between the Log and the *Discourse* reflect Ramusio's hand as editor – altering the text to emphasize the voyage's success as a means of promoting future voyages, and of course, future writings about them. Indeed, whoever wrote the *Discourse of the Great Captain from Dieppe* employed the Sumatran voyage as a kind of exotic lure that would appeal to readers, potential investors, and merchant-explorers with the aim of implicating them in future maritime expeditions made for profit, God, and country. Crignon's earlier account of faraway and exotic peoples who were mean, stingy, unreliable, heretical, and all-round bad people, would hardly inspire such acts of replication. Nor would the erratic actions of his friend and captain, or the meager cargo with which they returned.

Beyond the putative success of the voyage to Taprobana, the *Great Captain's Discourse* was a source of important and detailed intelligence about the locations, lands, products, and peoples of faraway lands. Crignon's Log similarly provides detailed information about each and every step of their voyage, noting the height of the sun nearly every day while at sea, and offering frequent estimations of longitude, details of wind directions, weather, the location of reefs and the depth of the seas. On land, and at every point where they do land, he notes, like the Great Captain, details about local foodstuffs and livestock, the terrain, the fauna and flora, the likelihood of finding gold, silver or spice, the customs of various peoples, their numbers, their clothing, the kinds of weapons they use, their weights and measures, and his impressions as to their reliability as potential trading partners.

But despite these similarities, the *Great Captain's Discourse* differs fundamentally from the Log by explicitly aiming to justify French incursions into waters and lands claimed by the Portuguese; it does this by intermingling the interests of France's merchant communities with those of the crown, arguing in strongly "nationalist" terms that the French – owing to their superior culture, piety, and commercial acumen – were uniquely suited to challenge Portuguese commercial and maritime hegemony in the Atlantic and beyond. This emphasis speaks not only to the difference between a ship's log and a retrospective compilation of various voyages contained in the *Discourse*, but to the span of time separating their writing, with the narrative arc of the *Discourse* clearly reflecting the escalating trade war between Ango and the Portuguese in the 1530s. For every location detailed in his account, Newfoundland, Norumbega, Brazil, Guinea, and Taprobana, the Great Captain illustrates and argues for the commercial rights of French merchants, justifying these by the discoveries they had made (of Terre Neuve by Jean Denys of Honfleur and Jean Ango, Sr.,[70] of Norumbega by Giovanni Verrazano,[71] and of parts of Brazil by Dionysius de Onfleur);[72] and where these were lacking, by arguing that the Portuguese had not built upon their discoveries by establishing bases, forts, castles, or indeed, amicable trading relations with the natives. As he says about the people of Brazil, "The people of this coast are much friendlier and more tractable with the French than with the Portuguese";[73] even the people of Guinea, he argues, who had been trading with the Portuguese for years, were "very happy when the French visit[ed] them."[74] The success of the Sumatran expedition did not stand alone, but was linked to a much more expansive view of France's role in the world. The agenda was clear and the Great Captain sets it out explicitly:

If I were asked why the Portuguese try to prevent the French from going to the land of Brazil and to such other lands as Guinea and Taprobana, I could give no other reason than that their insatiable avarice induces them to do this. Although the Portuguese are the smallest country on earth, no land is big enough to satisfy their cupidity. It seems to them that they hold in their closed fist that which they can never hold in both hands. I think that they have persuaded themselves that God created the sea and land for them alone, and that other countries are not worthy of navigating. If it were in their power to close the sea from Cape Finis[terre] to Ireland they would have done so long ago. It is as reasonable to stop the French from going to those lands where the Portuguese did not plant the Christian faith and in which they are not loved or obeyed, as it would be for us to stop the Portuguese from going to Scotland, Denmark, and Norway had we been there before them. Such a conquest is too easy and cheap, requiring no effort and encountering no resistance. The Portuguese are fortunate that the King of France is so kind and polite to them, for if he wished to unleash the merchants of his country they would conquer the markets and the friendship of the people of all the new countries in four or five years. They would do this with love and not by force, and they would penetrate further into these lands than the Portuguese were able to do in fifty years. Then the people of these lands would turn against the Portuguese as their worst enemies. This is one of the principal reasons why they do not want the French to talk to the natives, for after the French have had dealings with them the natives no longer wish to see the Portuguese, but hold them in great contempt and hatred.[75]

Crignon's persistent worries about the impiety of commerce, and his – and his Captain's – poetic efforts to intermingle religious devotion with

---

[70] Hoffman, 15.
[71] Ibid.
[72] Ibid., 22.
[73] Ibid., 21.
[74] Ibid., 26.
[75] Ibid., 23–24.

maritime exploration and trade as a means of quelling their anxieties, were here redirected away from their own commercial endeavors to those of the Portuguese. The Great Captain thus argued that rather than "being good Christians" and holding "God before their eyes" (as the French presumably did), the Portuguese looked only to "their profits."[76]

Halting Replication

Parmentier's voyage and those that followed surely did not look like spiritual voyages to the Portuguese; rather they seemed more like the actions of interlopers and pirates. As a consequence, the Portuguese did everything they could, whether through espionage, diplomacy or military force, to undermine Ango's commercial ambitions in lands and waters that they had claimed. Pirates and spies, explorers and the dynastic ambitions of kings converged on the ports of Normandy from the 1530s. Ango's pirates, translated into privateers by the king's letter of mark of July 26, 1530, pursued Portuguese caravels returning from the Indes, Guinée, and Brésil. But the letter of mark was just a formality, giving for a time the gloss of legality to an on-going free-for-all on the high seas (as we saw in Chapter 1).[77] After 1530, and the disastrous voyage of Le Sacre and La Pensée, the long-distance voyages sponsored by Ango and other French armateurs were directed to Guinée, Brésil, and Terre Neuve,[78] not to the East. Perhaps piracy was just more profitable than launching expeditions across increasingly well-patrolled trade routes to the Indes orientales.

Drifting Names

Explicit references to Parmentier all but disappear in the years immediately following the return of Le Sacre and La Pensée. Recall that the Great Captain's *Discourse* was anonymous and it was not until the

---

[76] Ibid., 23.
[77] Charles-André Julien, *Les débuts de l'expansion et de la colonization françaises* (Paris, 1947), 72 n. 5. The Portuguese responded by sinking or capturing Norman vessels, by attempting to control the makers of maps and the fruits of their efforts, and by pursuing diplomatic solutions (and exploiting dynastic connections) to stop Ango, and other similarly ambitious New Men. Where these strategies failed, they pursued other means – namely, bribery and naval force. Insofar as possible, the French king distanced himself from these affairs, alternatively acceding to, resisting, or ignoring the entreaties, demands and threats coming from all sides; sometimes he backed "his own" merchants, and sometimes he sold them out to achieve what he deemed more desirable ends (which usually meant besting Charles V of Spain in one way or another). For a detailed discussion of these complicated interactions, see Wintroub, *A Savage Mirror*, 21–38.
[78] See, for example, Mollat, *Le commerce*, 249–267.

nineteenth century, and the rediscovery of Crignon's Log, that it was related to Parmentier's voyage. The only explicit evidence for his voyage remained the little book published by Crignon in 1531. An unlooked-for trace, however, emerged far removed from the course he traveled and the lands he explored. Thus, Crignon's naming, in his elegiac poem, of the seas off Sumatra as the "Sea of Parmentiere" found a place decades later on an atlas by Diogo Homem (1558) on the other side of the world, where it appeared as the *Mare Le Parama(n)tiu(m)* above North America. Homem's map thus stretches and narrows America diagonally, placing an enlarged sea above and around it, not the Pacific, but "Parmantium," embracing it from the north to the west, and to the east and on to the Spice islands, India, and China (see Figure 6.11).

Diogo's brother, André Homem, completed his planisphere the following year in Antwerp; it shows an even larger, more westward oriented North America, the northwestern part of which traces into the nothingness of "Pars Yncognita," whilst still letting the east coast wander horizontally eastward into the Atlantic, leaving a clearly articulated Northwest Passage above (see Figure 6.12). Beginning just west of Labrador on his atlas a great sea appears: the "*Mare Ynventum per Paramantiel*."[79]

Diogo Homem, in all likelihood, placed the "Mare Le Paramantium" on his map on the basis of intelligence provided by Norman informants and colleagues; the same could probably be said about his brother. Exiled from Portugal for homicide, Diogo was living in London from 1546 when there were over 60 French pilots and cosmographers in the service of Henry VIII;[80] these included men such as Jean Maillard, the poet-cosmographer from Dieppe who references Crignon in his *Description de tous les portz de mer de l'univers* as one of France's greatest navigators;[81] Nicolas de Nicolay, who was to become royal cosmographer to Henri II in 1556; the Dieppois pilot Jean Ribaut who later led French expeditions to Florida in 1562 and 1565; Jean Rotz, of course; and Raulin Secalart (Raoullin le Taillois) of Honfleur, an associate of Jean Alphonse (a.k.a. Jean Fonteneau *dit* Alfonse de Saintonge) who participated in Roberval's expedition to Canada in 1542.

In addition to these maritime luminaries, one could, with reason, speculate that there were any number of sailors in England who had made the journey on fishing vessels sailing from France to Terre Neuve.

---

[79] See Wallis, *The Maps*, 66.
[80] On Homem's presence in London in 1546–7 see John Blake, "New Light on Diogo Homem, Portuguese Cartographer," *The Mariner's Mirror* 28:2 (1942): 148–160; E. G. R. Taylor, *Tudor Geography: 1485–1583* (London, 1930), esp. 59–74.
[81] Maillard, *France*, fol. 3ʳ.

Figure 6.11 Detail of the Queen Mary Atlas, Diogo Homem (1558).

Figure 6.12 Detail of North America from the Planisphere of Andreas Homem (1559).

According to Gaffarel, 60 Norman ships were armed to fish cod there in 1542; while from 1543 to 1545, in just the months of January and February, an average of two ships a day departed for Terre Neuve from the ports of Rouen, Dieppe, and Honfleur.[82] Perhaps among them were veterans of voyages by Parmentier and Crignon, or by Verrazano, Cartier, and Roberval, all of whom were charged with searching out a Northwest passage that would mirror Magellan's Straights in the South, something alluded to on the globe by Gemma Frisius and his student Gerard Mercator (c. 1537), where it was called *"Fretum arcticum sive Fretum Trium Fratrum"* (Arctic Strait or Strait of Three Brothers)[83] or on the atlas published by Frisius (1544) where a passage leading to Asia can clearly be found (see Figure 6.13).[84]

---

[82] Not including those from Brittany or the Basque region. Paul Gaffarel, *Ango*, 18; see also Gosselin, *Documents authentiques*, 12. See Laurier Turgeon, "Codfish, Consumption, and Colonization: The Creation of the French Atlantic World During the Sixteenth Century," in Caroline A. Williams (ed.), *Bridging the Early Modern Atlantic World: People, Products, and Practices on the Move* (Surrey and Burlington, 2009), 33–56, at 36–37.

[83] See, for example, Alan Day, *Historical Dictionary of the Discovery and Exploration of the Northwest Passage* (Lanham, Maryland, 2006), 176.

[84] To the north of Baccalearum (from *baccalo*, Portuguese for cod), "terra florida" in the Homem atlases. The nomenclature Insula Baccalauras and Terra Nova were first employed by Johannes Ruysch in *Universalior Cogniti Orbis Tabula, Ex recentibus confecta observationibus* (1508).

Frisius's map was not entirely dissimilar to Sebastian Münster's *Geographia*, first published in 1540, which identified the Northeastern part of the New World as *Francisca* and indicates with a legend that the strait above it is "the route to the Moluccas..." (*Per hoc fretu iter patet ad Molucas*) (see Figure 6.14).[85]

The maps by the Homem brothers conform to the general outline supplied by Frisius's later map, the *Novae Insulae XVII*, which eliminates the legend, and intimates the possibility of a passage obscured by the placement of a black ink frame. Perhaps, following Münster's (or Fine's similar) identification of a New World Francisca, and being well acquainted with informants who identified Newfoundland (*Terre Neuve*) and Labrador as French possessions, they (the Homems) named the sea bearing the northwest passage after Parmentier.

Abraham Ortelius, on his well known 1566 world map, following on and refining the work of the Homems, translated (*relevé*) this sea into a cape – *Paramantia*, the northernmost point of the region of America that he labeled Nuova Franza on his map (see Figure 6.15).

Humphrey Gilbert, half-brother to Walter Raleigh, soldier of fortune, member of Parliament, and maritime explorer, follows Ortelius closely, specifically referencing Cape Paramentia on both his map and in his treatise "Voyages in Search of the North-West Passage" (see Figure 6.16).[86]

Gilbert's map was made with the explicit intention of drumming up enthusiasm and investors for a proposed series of voyages to search for the Northwest Passage. His map, like those that came before it in referencing Parmentier, clearly demonstrates the reason for the drift of Parmentier's name from the Indian Ocean into the north Atlantic, showing the place it marked to be part of passage from east to west above (a now much wider) America (linked on its western extremes with Asia). Parmentier was thus a stand-in, a place holder, for the Pacific; it then became a southern cape along the northwest passage to the Moluccas; finally, it disappeared altogether, to be replaced by Frobisher who reasoned that "his discovery," like that of Magellan's in the south, ought to bear his own name – "The Frobisher Straits," which were, as indicated on George Best's map of 1578, "*The Way Trendin to Cathia*" (see Figure 6.17).[87]

---

[85] The legend is removed from later editions; see Surekha Davies, "America and Amerindians in Sebastian Munster's Cosmographiae universalis libri VI (1550)," *Renaissance Studies* 25:3 (2011): 351–373, at 366; the 1540 map was, she says, reprinted in the "1541, 1542 and 1545 editions of Ptolemy's *Geographia*, and in the 1544, 1545, 1546 and 1548 editions of Münster's *Cosmographia*."

[86] Written in 1566, but published in 1576.

[87] George Best, *A true discourse of the three Voyages of discouerie, for the finding of a passage to Cathaya, by the Northwest, vnder the conduct of Martin Frobisher Generall* in Richard Hakluyt and C. R. Beazley (eds.), *Voyages of Hawkins, Frobisher and Drake: Select Narratives from the 'Principal Navigations' of Hakluyt* (Google eBook), 91.

Figure 6.13 Detail of the Atlas published with Petrus Apianus's *Cosmographia* by Gemma Frisius (1544).

Figure 6.14 Detail of a map by Sebastian Munster, *Typus orbis universalis* (Basal, 1545).

Frobisher's expedition to look for the Northwest Passage began in June 1576. He found his eponymous straits leading to Cathay in early August. When he returned to England in October he brought back with him a sample of rock as a "token" of the "Christian possession" of the lands he found. When this was assayed by the Venetian Alchemist Giovanni Battista Agnello it was found to contain a good amount of gold.[88] In addition to the rock, Frobisher also took a human "token," a native prisoner, as proof and "witness of the captain's far and tedious travel" undertaken in search after the long sought transit to the East.[89] The prisoner died promptly after reaching London, but he, and the black rock, did their jobs, and another expedition was organized for the following year.[90]

---

[88] The story here is complicated; Agnello provided different accounts of the amount of gold contained in the rock. Originally, he estimated that it was worth £30 per ton (a not insignificant amount); he revised this upward when speaking to potential investors, estimating its worth at 240 per ton, "which would pay a return of 2,400 percent on expenses." Ibid., 94. Regarding Agnello's assay see Deborah Harkness, "Strange Ideas and English Knowledge," in Smith and Findlen, *Merchants and Marvels*, 137–162, at 152–154.

[89] Ibid., 93.

[90] See R. McGhee, *Arctic Voyages of Martin Frobisher: An Elizabethan Venture* (Montreal and London, 2001), 63.

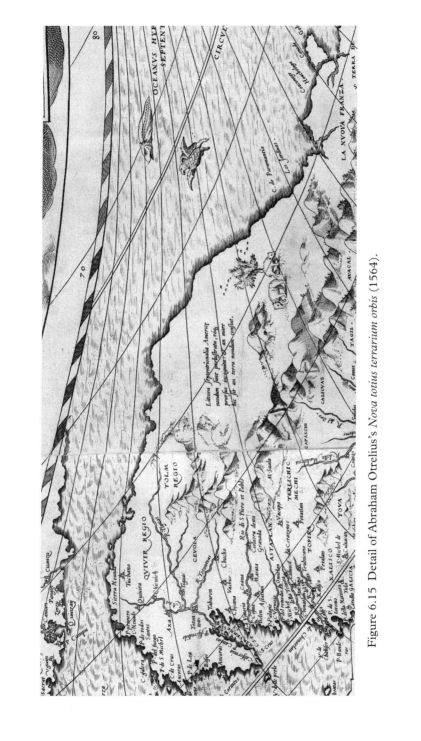

Figure 6.15 Detail of Abraham Ortelius's *Nova totius terrarium orbis* (1564).

Figure 6.16 Detail of Humphrey Gilbert's world map (1576).

The promise of gold was greeted with enthusiasm by potential investors, and it clearly outweighed the search for a Northwest Passage in
planning Frobisher's second expedition. As George Best put this in his
True Discourse of [Frobisher's] Three Voyages: "the hope of more of the same
gold ore to be found kindled a greater opinion in the hearts of many to
advance the voyage again. Whereupon preparation was made for a new
voyage against the year following, and the captain more specially directed
by commission for the search more of this gold ore than for the searching any further discovery of the passage."[91] Three ships set sail the following year; they returned in October with two hundred tons of rocks.
The Queen appointed a special commission to evaluate the results; they
determined, "after sufficient trial and proof made of the ore, and having
understood by sundry reasons and substantial grounds, the possibility
and likelihood of the passage, advertised her Highness that the cause was
of importance, and the voyage greatly worthy to be advanced again."[92]
The next year a major expedition was launched with fifteen ships. The
search for the Northwest Passage, which Frobisher was convinced he had
found, was thus augmented, and to a certain extent, side-tracked, by a
hunt for gold. Over a thousand tons of rock returned to London in 1578.
The truth, however, soon became apparent.[93] The initial reports were
wrong; Frobisher was wrong; the assayers were wrong; the commission

[91] Ibid., 95.
[92] Ibid., 134, and 90–92.
[93] Ibid., 140–142.

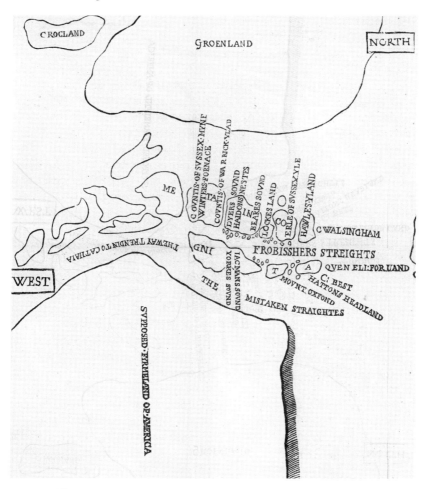

Figure 6.17 George Beste, *A true discourse of the late voyages of discouerie,*
for the finding of a passage to Cathaya, by the Northweast, vnder the conduct
of Martin Frobisher Generall, etc. (London, 1578).

set up by the Queen was wrong: the rocks were worthless, the investors
had lost everything, and the passageway to the east was nothing but a
hoped for and enigmatic dream.

From such mistakes and failures maps were made. Indeed, Parmentier's
voyage spawned mutations, luring voyagers onto new seas and to new
lands, but still voyages like his own: qualified failures, transformed iterations, continuations, extensions, replies and replications, that had the
power to move ships, men, knowledge and goods back and forth across

wide expanses of water. Thus was the name given to a sea off the coast of Sumatra by an obscure Norman poet transformed into part of a mythic Northwest Passage depicted with cartographic surety on maps by the Homems, Gilbert and Ortelius to demarcate an imaginary space of passage – a hope, an expectation, a desire built from accretions of previous knowledge, techniques, older maps, traces and mis-readings, improvisations, real and rumored voyages, tall tales, and myths. Parmentier's name, borrowed from a poem, or perhaps lent by an anonymous sailor, marked an early stage in this adventure, this search, for a passage west, not only as a signpost marking the way, but as part of a concerted effort to implicate, enfold, entwine, and entangle men into investing in and making future voyages across the seas in search of profit, glory and God.

Thus Parmentier's name, before it was forgotten, was a referral, a reference, a place on the way and, as such, a supplication, an entreaty, a goad, a trap, a lure – *implicare*: to be "entangled," to be "implicated" or "inducted," but also to be "confused" and "befuddled," brought into "the briars,"[94] amongst islands, islets, capes, ice floes, bays and dead ends in places that were difficult – and indeed, often impossible – to navigate, and almost too cold even to dream about, but which promised untold riches, adventure, spice, gold, converts and glory, truly a *Meta Incognita* as Queen Elizabeth named the lands Frobisher had discovered. The way was not clear, as evinced by Best's addition of the "Mistaken Straits" to his map, but the hope marked, if but temporarily, by a forgotten name, held a promise even in its absence: "The Way Trendin Towards Cathia."

Parmentiere, Paramantiam, Paramantiel, Paramantia – like the half hidden place denoting Taprobana on Fine's map – was a possession marked by its absence, a possibility of presence to be strived for and mathematically positioned (*relevé*) in an imaginary geography that was there and then was not: a (non)place that would carry minds, men and ships across the world to glory and to profit, not so much as an escale or a destination, but as a reference, a guide, a marker, a means of persuasion, an induction – a possibility that might replicate future merchant-explorers, investors, cartographers, maps and voyages.

The Discovery of the Age of Discovery, or New Uses for Old Names

Bodies of knowledge, techniques, rules and motivations do not arise *sui generis*, but are part of complex, continuous, growing, evolving, and often contradictory, traditions, survivals, artifacts and associations that are just as

---

[94] See OED, s.v. implicate; see also Cotgrave, *A Dictionarie*, s.v. *impliquer*.

often lost from sight as they are named, written about, or consciously relied upon. Though Parmentier more or less disappeared from view after the Pensée and the Sacre returned to Dieppe, a kind of "homeopathic" remainder of his voyage continued to act as a ghost in the machine of transoceanic commerce. To paraphrase Pierre Nora's apt description, Parmentier's memory survived as a trace taking "refuge in gestures and habits, in skills passed down by unspoken traditions, in the body's inherent self-knowledge, in unstudied reflexes and ingrained memories, and memory transformed by its passage through history."[95] Crignon's Log, his poetry, the Great Captain's *Discourse*, the veteran sailors who successfully completed the voyage, the "savages" they brought back from the island of Saint Hélène, all contributed in one way or another to future voyages and voyagers, future texts, future maps, future instruments, future cross-cultural encounters, future shipboard relations, and the future desires and motivations of investors, sailors, merchants and kings. One can even see in the voyages' utter failure, a *replique* – a response – in the abandonment of the East for the West, that is, in the attempts by the French to colonize Newfoundland (1534), Brazil (1555) and Florida (1562), and the almost complete absence of future voyages to the Indies for decades.

Though there was a sort of subterranean continuity entangling Parmentier in the *longue durée* of maritime exploration and trade, on the surface he had become little more than a shadow; indeed, for centuries all that explicitly remained of him were brief, often error strewn, anecdotes about his expedition to the Moluccas (sic) and to China (sic) in local histories, biographical dictionaries, and in works on the history of theater and poetry.[96] It wasn't until the nineteenth century that Parmentier again emerged into full relief, with blood and bones, ships and men, in texts revived to serve a new purpose. Thus did memory once again intertwine with histories, rather than poesies, written to inspire and persuade new generations to cross the seas for profit, truth and glory. Indeed, just as Jean Parmentier was translated from the seas off Sumatra to the Northernmost reaches of the Americas in the late sixteenth century, now he was translated across the centuries to become a nationalist emblem of renascent French colonial ambition.

Beginning in the 1830s, shortly after France's invasion of Algeria, Parmentier took on this new position, no longer a placeholder to a

[95] Pierre Nora, "Between Memory and History: Les Lieux de Mémoire," *Representations* 26 (Spring, 1989): 7–24, at 13.

[96] For example, Jean-Antoine-Samson Desmarquets, *Mémoires chronologiques pour servir à l'histoire de Dieppe…* 2 Vols (Dieppe, 1785), I: 112, and II: 10; M. de Beauchamps, *Recherches sur les théâtres de France* (Paris, 1735), 312–314; P. G. Le Mercier, *Histoire du théâtre François, depuis son origine jusqu'à présent* (Paris, 1745), 265–268.

Northwest Passage, but a passage himself – a transit to the imperial expansion of France throughout the world, and a historical exemplar of Gallic heroism, daring, and ingenuity. This recuperation was an almost archaeological process, digging beyond the debris of myths and anecdotes back to a foundation in the sources. The text of the log of Parmentier's voyage was first published by Louis Estancelin, a Norman historian, archivist, and parliamentarian, in his *Recherches sur les voyages et découvertes des navigateurs normands* in 1832. The following year, Ludovic Vitet, architect, politician, and *Inspecteur général des Monuments historiques*, explained – in introducing his republication of the Log – that Estancelin discovered this "monument" of Dieppe's glorious history while visiting his friend, the printer-bookseller and local historian M. Théodore Tarbé in Sens.[97] Tarbe, Vitet tells us, "inherited this manuscript from one of his brothers, a merchant from Rouen who made the history of commerce and navigation the object of serious study. The manuscript was entitled *Voyage des Dieppois*, and Estancelin, Vitet tells us, instantly recognized it as an account of Parmentier's 1529 voyage to Sumatra."[98] In 1846, Léon Guérin, a naval historian and popular children's writer, included a lengthy account of Parmentier's voyage in his *Les Navigateurs français: histoire des navigations, découvertes et colonisations françaises*.[99] His sources were Estancelin's publication of the Log, supplemented by the *Discourse of the Great Captain*, and Crignon's 1531 collection published by Morrhy. Some 20 years later, relying on the same sources, though more deeply mining Parmentier's poetry, Pierre Margry, an historian and archivist at the Ministère de la Marine et des Colonies, provided an account of the voyage in his *Navigations françaises et La Revolution maritime*.[100] In 1883, the renowned Orientalist Charles Schefer published a more complete version of the Log included in an early eighteenth-century manuscript copy entitled *Le discours de la navigation de Jean et Raoul Parmentier, de Dieppe*.[101] That same year, Margry produced a critical edition of the Log published by Estancelin in 1832, entitled *Journal d'une navigation des Dieppois dans les mers orientales*. Subsequently, accounts appeared in a number of sources, most notably

---

[97] Ludovic Vitet, *Histoire des anciennes villes de France: Dieppe* (Paris, 1833), 67–111.

[98] Ibid., 68. This identification also allowed Estancelin to put a name – or at least hone in on this voyage – as the one described by the Great Captain of Dieppe published by Ramusio.

[99] Léon Guérin, *Les navigateurs français: histoire des navigations, découvertes et colonisations françaises* (Paris, 1846), 119–161.

[100] Pierre Margry, *Les navigations françaises et la révolution maritime du XIVe au XVIe siècle* (Paris, 1867), passim.

[101] This is a transcription of BN, n.a. fr. 7510.

in Paul Gaffarel's *Jean Ango* (Rouen, 1889),[102] and in Eugène Guénin's *Ango es ses pilotes* (Paris, 1901).[103]

Though the particular political and social circumstance framing the work of these authors, spanning approximately three generations, differed considerably, all were concerned with promoting the expansion of French colonial power. Louis Estancelin (1777–1858), for example, was an unrelenting and unapologetic proponent of colonization, not so much as part of a mission to bring "civilization" to the "savages," but as a kind of safety valve which would help ameliorate the social and political tensions concomitant with processes of modernization at home. For him, the "savages" of the most concern were not indigenous peoples abroad, but workers in France whose traditional ways of life had been undermined by scientific and technical progress. Estancelin was by no means a romantic technophobe; indeed, modernization, for him, was a patriotic necessity. However, the glorious march of progress into the future that he envisaged needed to be accompanied by the expansion of French power into the world. As he succinctly put it, there were only two alternatives, either "colonization or revolution."[104]

Pierre Margry (1818–94), on the other hand, was less concerned with colonization as a solution to the social dislocations wrought by "modernization" at home than he was with promoting the glory and economic power of France in a competitive world system. A long-time archivist at the Ministère de la Marine et des Colonies, he explicitly compiled his collections of France's maritime history as a means of "rendering justice to the role that our country has had in the far flung regions of the world."[105] As he put it, "in founding far away colonies, [France] not only accrued for herself riches and power, but opened up vast fields to her activity thus extending the reach of her glory (*domaine de gloire*)."[106] He concluded by noting that Parmentier's voyage could not be forgotten for two reasons: the first, he said, is "purely scientific, belonging to the history of the progress of geography"; the second, he continues, is "essentially political, and linked with the development of the liberties of the world and, as such, to the history of the Law of Nations (*Les Droits des Gens*)."[107]

---

[102] Paul Gaffarel, *Jean Ango* (Rouen, 1889), 41–51.
[103] Guénin, *Ango es ses pilotes*, 126–144. He also provides an account of Parmentier's voyage in his *La route de l'Inde* (Paris, 1903), 136–153.
[104] Louis Estancelin, *Études sur l'état actuel de la marine et des colonies françaises* (Paris, 1849), 172–173.
[105] Margry, *Les navigations*, 352.
[106] Pierre Margry, *Familles de la France colonial* (Paris, 1851), 3.
[107] Margry, *Les navigations*, 220.

Charles Schefer (1820–98), who published the manuscript of the Log now at the Bibliothèque Nationale, was an adventurer, linguist, diplomat, and collector, perhaps best known in his day as the most prominent Orientalist of his generation. In addition to holding a chair at – and then acting as president of – the Ecole des Langues Orientales, he was a prolific writer and editor of texts on travel, history, and geography. His life as a scholar was supplemented by work in the field. In the late 1840s and early 1850s he was a contributing foreign correspondent to the Société de Géographie in Turkey. In 1851, for example, he obtained several maps of the provinces of the Ottoman Empire for the Société; he announced that authorization had been obtained for archeological excavations in the area of "Mossoul" and "Pachalik de Bagdad," and that he had prepared a study on the territorial divisions of the Ottoman Empire requested by Alexandre Dezos de la Roquette, the Secrétaire general de la commission central of the Société and editor-in-chief of its principle publication, the *Bulletin de la Société de Géographie*.[108] Throughout the 1860s and through the 1880s, he also worked on diplomatic missions for the Département des affaires étrangères. In 1862, for instance, he negotiated the sale of a naval base located in the Gulf of Aden, which formed the origin of the French colony at the mouth of the Red Sea.[109] According to Paul Gaffarel, writing in 1879, this colony had the potential to be a veritable *"embryon d'une nouvelle France."*[110] It would, he said, be "one of the great ports of the Indian Ocean. Would that our wish for prosperity be some day realized!"[111] The unbridled enthusiasm of Gaffarel for French imperial expansion contrasts with the more subdued back stage activities of Schefer. Gaffarel (1843–1920) was a tireless and vocal booster of French colonialism, viewing it not only as a consolation for France's recent defeat by the Prussians, but as part of a life-and-death struggle that would lead either to scientific, economic, and moral progress, or to stagnation and decline.

Born in Moulins, about as far away from an ocean as one can get in France, Gaffarel was to become one of France's most prominent historians of maritime exploration and transatlantic empire. Agrégé in history and geography in 1865, he did his doctorat at the École normale supérieure, writing a thesis on *Les rapports de l'Amérique et de l'ancien continent avant Christophe Colomb*. In addition to teaching history and geography

---

[108] *Actes de la Société. Procès-verbaux des séances*, … Presidence de M. Jomard, 19 décembre 1851, *Bulletin de la Société de Géographie* (July–December, 1951), t. 2: 461.

[109] See Auguste Bouché-Leclercq, *Notice sur la vie et les travaux de M. Charles Schefer* (Paris, 1899), 24–25.

[110] Paul Gaffarel, "Obock," *Société normande de Géographie* (1879), 123.

[111] Ibid., 126.

at the University of Dijon, he was an active member of the Société de Géographie in Paris, the founder and first secretary general of Dijon's Société Géographic, and a corresponding member of the geographical societies of Bordeaux, Lyons, Rouen, and Marseilles. Gaffarel was also a prolific writer about what he termed the *"génie colonisateur"* of France. In his view, metropolitan regeneration required colonial expansion. He saw this as part of a *"mission civilisatrice."* As he put it with regard to the "Far East" – "the natives are children who are just being admitted to civilization." He was clear about what course should be followed:

Is it not our duty to direct them, to instruct them, to educate them morally? In Cochin and China, as in Senegal or in Algeria, as everywhere that we find ourselves in the presence of primitive or corrupt societies, our most useful auxiliaries will be missionaries and schoolmasters. What force can resist the two levers of religion and science? Let us know how to use them and we shall have accomplished a useful and patriotic work.[112]

This civilizing mission would, he thought, also perform a patriotic service for the nation, warding off stagnation, promoting economic growth, and ameliorating social conflict.

Eugène Guénin (1865–1931) was similarly explicit about why he had undertaken to write about French maritime and colonial history, arguing in his *Histoire de la colonisation française* that the nation of France was particularly "suited to colonisation." He thus viewed his own historical work as a contribution to the creation of a new colonial empire of which the nation could be proud. He explains that at one time France "possessed a marvelous colonial empire in Canada and the Indies," but that it had been "lost" because of governmental inaction. But since then, France had taken her destiny back into her hands and conquered another empire. And having become again a great colonial power, it was important "to profit from teaching the past so as to not repeat the mistakes of another time."[113]

Though their reasons varied, as did the circumstances within which their work was produced and received, these "historians" revived and publicized Parmentier's lost voyage, making monuments – and establishing archives – of its memory. Vitet is explicit about this; abandoning

---

[112] Agnes Murphy, *The Ideology of French Imperialism, 1871–1881* (New York, 1948), 204–205. Gaffarel ideas closely resemble those of Scheffer's friend and colleague, Ernst Renan. See, for example, Jonathan Dewald, *Lost Worlds: The Emergence of French Social History, 1815–1970* (Philadelphia, 2010), 29. Surely thinkers such as Gaffarel and Renan built the scholarly foundation for colonial policies advocated by Jules Ferry. See, for example, Christopher Hill, *National History and the World of Nations: Capital, State, and the Rhetoric of History in Japan, France, and the United States* (Durham, 2008), 134ff.

[113] Eugène Guénin, *Histoire de la colonisation française* (Paris, 1898), 7–8.

his "conjectural" discussion of the Norman "discovery" of America in 1488 by the mysterious Dieppois navigator Jean Cousin, he argues for the return "to history and to facts for which we can supply proof."[114] Thus, he continues, "of all the monuments of which we can attest to the reality of the great feats of navigation accomplished by the Dieppois, the journal of Jean Parmentier is assuredly the most definitive and the most incontestable."[115]

Vitet was a romantic, but a kind of romantic positivist.[116] For him, Estancelin's discovery was the very stuff (the *res*) of history. It was proof – but proof that needed to be conjoined with *plicare* – to fold, entwine, and entangle, that is, to persuade (*replicare*). In other words, for him history was composed of exemplary monuments of a heroic past that were to be uncovered, restored, displayed and written about; these monuments were textual as well as architectural. Vitet's historiography was thus closely allied with his work as the Inspecteur général des monuments historiques. Like his account of the discovery and publication of Crignon's Log, he considered the material traces of France's past as "witnesses" to a sacred *patrimoine*.[117] As he said about the Church of Notre-Dame de Noyon, "we must interrogate the monument itself."[118] Thus Vitet located in the Church's structure – in particular, the style and use of its gothic arches – a kind of text revealing a "*movement tout nouveau des esprits*," "a spirit at once innovative, risky and systematic that could be traced to the first battles of reason against authority, of the nascent bourgeoisie against a feudalism in decline, of the popular and living language against an ancient priestly tongue that was in the process of dying."[119] The discovery and publication of Crignon's Log constituted for him precisely the sort of persuasive relic from the past that could testify to the emergence

---

[114] Vitet, *Histoire des anciennes villes*, 67.

[115] Ibid., 67–68.

[116] See, for example, John Tresch, *The Romantic Machine: Utopian Science and Technology after Napoleon* (Chicago, 2012).

[117] At the July 1845 séance of the Société, the necessity of extending these endeavors beyond metropolitan France to the "classement des monuments" in Algeria was discussed. See Françoise Berce, *Les premiers travaux de la commission des monuments historiques, 1837–1848* (Paris, 1979), 364. See also Goran Blix, *From Paris to Pompeii: French Romanticism and the Cultural Politics of Archaeology* (Philadelphia, 2013), 23, and Ludovic Vitet, *Etudes sur l'histoire de l'art* (Paris, 1868), 405–417. The connection between archaeology and historiography finds expression in Vitet's work with the Comité des monumens inédits de la litterature, de la philosophic, des sciences, et des arts. See, for example, Kevin D. Murphy, *Memory and Modernity: Viollet-le-Duc at Vezelay* (Philadelphia, 1999), 43ff.

[118] See Ludovic Vitet, *Monographie de l'église Notre-Dame de Noyon* (Paris, 1845), 3. See also Blix, *From Paris to Pompeii*, 125.

[119] See Blix, *From Paris to Pompeii*, 120. See also André Vauchez, "The Cathedral," in Pierre Nora, Lawrence D. Kritzman (eds.), *Realms of Memory: Traditions* (New York, 1967), 57.

of a similar *movement tout nouveau des esprits* in the early sixteenth century, thereby lending credence to his earlier conjectures that France had
indeed discovered America years before Columbus's voyage in 1492.[120]

The Norman geographer and historian, Gabriel Gravier, who wrote
about Parmentier in 1902, expressed views similar to Vitet, namely,
that France's history might supply the raw materials of *amor patriae*
and moral progress. As he put it: "I believed, and I still believe, that
the moral level of a people is elevated by the memory (the *souvenir*) of
great men."[121] He thus wrote copiously on France's colonial and maritime history, publishing books and writing articles on Cavelier de La
Salle, the Discovery of America by the Normans, Marquette's travels
on the Mississippi, Verrazano's voyage and Samuel Champlain's life.
Gravier, however, belonged to a different generation than Vitet. Whereas
Vitet sought to champion the past as a site of national pride through the
collecting, classification and restoration of its textual and material monuments in histories that would possess readers in the same way as a novel
by Victor Hugo (with whom he worked on the Commission des monuments historique), Gravier aimed to secure the same ends through the
institutional power of organizations such as the Société de Géographie,
the Société de Géographie et Commerce, and the Séction Dieppois de la
Société normande de Géographie, which he founded. These geographic
associations were, he argued,

endowed with the capacity to arouse, indeed create, explorers who would bring
together science, bravery, initiative and physical force. They direct them, support
them, give and find them backing, and make them, and their work, known to
the world. The earth and its peoples reveal themselves and deliver their secrets;
each day a piece of the map is filled in. The history of Geography will [one day]
say that the last thirty years ... have been the heroic age of the discovery of the
world. [...] The result of this long and admirable campaign has been the recon-
stitution of our colonial empire... This magnificent domain, where today our civilization, our commerce and our industry thrive is an extension (*un prolongement*)
of France.[122]

The revival and replication of Parmentier's voyage was thus part
of an agenda that was explicitly expansionist and imperial – directed
both in time and across space. The idea of France, translated into persuasive and patriotic narratives written about the recovery, restoration
and preservation of its textual and material patrimony, was paralleled

---

[120] One can see a similar romanticization of Normandy's merchant explorers, though from
a very different political perspective, in the play by Auguste Luchet and Félix Pyat,
Ango: drame en cinq actes, six tableaux, avec un epilogue (Paris, 1835).
[121] Gabriel Gravier, *Les Normands sur la route des Indes* (1880), 8.
[122] Ibid.

by the extension of France beyond her European frontiers. As Gaffarel said: "colonization begins with history."[123] Indeed, the "proof" supplied by such historical "monuments" as Crignon's Log, aimed to implicate and enfold publics and politicians in the cause of replicating France elsewhere. They were to do this first as historiographic method and then as institutional imperative. Recall Cotgrave's definition of "induction" as "an inducement, allurement, or persuasion, unto" and also, "a forme of argument from particulars to universals." Here peripheries and faraway lands were literally to become extensions (*prolongement*) of the metropole, which was to be replicated far and wide; or as Gaffarel desired, a reproduction of "New Frances" around the globe: "an African France, an Oriental France, an Equinoxial France, an Oceanic France."[124] Parmentier thus became the material, the *res*, and the persuasive point of reference, justifying and validating the replication of new colonial empires and new regimes of universal knowledge.

Indeed, the recuperation of France's "Age of Discovery," its collection as primary source material – in edited volumes, in histories of geography and in maritime histories – grew alongside the emergence of ethnographic collections,[125] the documenting of the languages, the architecture and the monuments of France and her extended empire, and the development of academic disciplines of – and institutions representing – geography, ethnology, history, and archaeology. Though Parmentier's voyage failed, it nevertheless lived on as a monument to an inaugural founding moment of expansionary power.

---

[123] Paul Gaffarel, *Les colonies françaises* (Paris, 1880), 1.

[124] Ibid., 8–15. At 120, 170, 180, and 380 as cited by Murphy, *The Ideology*, 194.

[125] Such as those advocated by Edme-François Jomard (editor of the *Description de L'Égypte*, member of the Institut d'Egypt established by Napoleon, founding member of the Société de Géographie de Paris, and member, along with Vitet, of the Commission des antiquités de la France) who saw "the assembling and methodical classification of artifacts as a primary function of an emerging science of "ethnography." See Elizabeth A. Williams, "Art and Artifact at the Trocadero," in George Stocking, Jr., *Objects and Others: Essays on Museums and Material Culture* (Madison, 1988), 149.

Epilogue: Pirate Epistemologies

"Who are you, stranger? From where have you set sail? Along liquid
paths? Do you roam for trade?

Or for adventure, crossing the seas, like pirates,

Risking their lives and bringing harm to others?"

– Homer[1]

One last word. Pirate.

Pirate, of course, was a relative term; what for one was a pirate, for
another was a privateer or a royal servant. The Latin, *pirate*, for sailor,
mercenary, armed robber or brigand, was a transliteration of its Greek
etymon, *peirates*. The Greeks employed the word in ways similar to the
Romans – as sea dogs and ocean-going villains. In both cases, however, there were stowaway associations, e.g. *peira* – to "try, attempt, or
attack": to put to "proof" – and *empeiros*, "to be experienced and to
experience," as well as "to be skilled and expert" (*empeiria*).[2] These
notions – and the practices associated with them – found their way into
Latin as *experiri* and *expertus*. We are only a step away from *empirique*,
experience and *expertise*": "sea dogs" on the hunt.[3] This is perhaps why
experts were – and are – so mistrusted: deep down they are shifty, crafty,
and cunning; they are pirates.

Experience was, as we have seen, knowledge acquired "by much practise, and many trialls;"[4] this knowledge was closely akin to experiment,
which finds an additional self-validating meaning as *preuve*, from the latin

---

[1] Quoted as the epigram of Daniel Heller-Roazen's *The Enemy of All: Piracy and the Law of
Nations* (New York, 2009).
[2] See ibid., 35 and OED, s.v. pirate. I would like to thank Roland Greene for pointing this etymology out to me. See Ronald Sousa, "Cannibal, Cartographer, Soldier,
Spy: The Peirai of Mendes Pinto's *Peregrinação*," in Elizabeth Fowler and Roland Greene
(eds.), *The Project of Prose in Early Modern Europe and the New World* (Cambridge, UK,
1997), 15–30.
[3] Recall the history of "information" that links it to dogs barking at the scent of prey. See
Chapter 1.
[4] Cotgrave, *A Dictionarie*, s.v. *experience*.

probare, which for Estienne becomes the adjective, *probus*, and following Cicero *"un bon preudhomme,"* a *"Homme de bien,"* and Plautus, *"bonne et loyale marchandise."*[5] Truth-speaking, authenticating evidence, and social standing were thus entangled with success and status, that is with *prou*'s meanings as "noble," "honorable," and "gallant"; they were also connected to its meanings as "profit" and "advantage."[6] We have come a long way from pirates it seems, but not so far that profit wasn't seen as the measure of success, either as proved veracity (*un bon preudhomme"*) or reliable (*"bonne et loyale"*) merchandise.

These social, material, and ethical measures of proof shade into the pejorative, however, when experience becomes practice reliant on observation, bodily knowledge, and material engagement with the world (*empiric*). For Aristotle, *empeiria* stood between perception and understanding. Though propaedeutic to the latter, observational knowledge was clearly inferior to it; indeed, it was not really knowledge at all, but raw data divorced from the disciplines, institutions, and philosophical systems that could provide it with meaning.[7] This inferiority was challenged by the so-called empirical school of Greek physicians who in contrast to Galen derived their therapies from observing, collecting, and organizing the data of experience into case histories.[8] From the perspective of Galen and his followers, empirics personified Aristotelian critiques of non-theoretical expertise as menial, ignorant, cunning, and duplicitous. Labeled as charlatans and quacks, this negative view of empirics became commonplace and survived well into the eighteenth century.[9] The fourteenth-century physician, Guy de Chauliac, sums up the commonly held view that empirics (*empiriques*) were "lewde men." Elsewhere, he referred to them as men without academic training who *"experimentez & incantacionz.i. charmez,"* thus bringing the scope of empirical expertise into the region of cunning women and witches who employed incantations and charms to dupe the credulous.[10] De Chauliac's words mirrored contemporaneous institutional-disciplinary sanctions and the instigation of trials of unlicensed empirics by the faculty of medicine at the University of Paris.[11] At the same time, the shifting social and

[5] Estienne, *Dictionarium*, s.v. *probus*.
[6] Jessie Crosland, "Prou, Preux, Preux Hom, Preud'ome," *French Studies* (1947): 149–156.
[7] See, for example, Travis Butler, "Empeiria in Aristotle," *The Southern Journal of Philosophy* 41 (2003), 329–350.
[8] As for example, Parmentier's shipboard postmortem exams (see Chapter 3).
[9] See, for example, Stephen Pumfrey, "Who Did the Work? Experimental Philosophers and Public Demonstrators in Augustan England," *British Journal for the History of Science* 28:2 (1995): 131–156.
[10] OED, s.v. empiric. See, also Eamon's *Science and the Secrets of Nature*, 194–233.
[11] See, for example, Pamela Smith, *The Body of the Artisan*, 158–159.

commercial terrain in late medieval and early modern Europe gave impetus to claims – made with increasing frequency and backed up by promises of utility and profit – that knowledge and status could derive from active engagement with the world of particulars. There were nevertheless strong (derogating social and epistemological) impediments to forging fast and sure connections between empirics and the *preud'homme*.[12] In the last quarter of the sixteenth century, for example, André Du Breil, a physician at the University of Paris, lumped empirics into the same bag as "vagabonds, atheists, exiles, priests, monks, shoemakers, carders, drapers, weavers, masons, madams and prostitutes."[13] How, then, could piracy (*peirates*) and the hunt for intelligence (*empeiros*) be transformed into pious quests for profit and credible knowledge? Could expertise ever be disentangled from craftiness; translation from treason; facts from the factitious; induction from cunning; evidence from the plebian; profit from piracy; commerce from ungodly acquisitiveness; and replication from acts of practised rehearsal? The question, of course, is rhetorical, and the answer surely involves a shift of focus from the theoretical mapping out of the so-called Great Divide to an historical examination of the practices, technologies, and dispositions by which new men such as Ango, Parmentier, and Crignon attempted to navigate the social, epistemic, and spiritual geographies of their world. The modern, in this sense, was not tilted to one side or the other in a binary balanced out by a primitive pre-modern other; rather, it was composed of the human material of pirates, traders, savages, artisans, sailors, poets, experts, and priests entangled with each other, and with the things that they hunted for: that is, with booty, fetishes, works of art, intelligence, goods, instruments, sacramental objects, converts, and coin. Indeed, empirics were hunting for more than just knowledge; their hunt was also for profit and power. Taking the height of the sun every day at noon with an astrolabe, performing postmortem examinations, and sailing ships, were all part of this hunt, but so too was taking and executing Minangkabau hostages, trading for pepper, and writing Marian verse.

As the voyage of Le Sacre and La Pensée demonstrates, the divides of geographical, temporal, and social space were interpenetrating and discontinuous, surging with reversals and rocking with inflected continuities,

---

[12] See especially Edgar Zilsel, *The Social Origins*.

[13] Quoted in Alison Klairmont Lingo, "Empirics and Charlatans in Early Modern France: The Genesis of the Classification of the 'Other' in Medical Practice," *Journal of Social History*, 19:4 (1986): 83–603; see also, James Farr, *The Work of France: Labor and Culture in Early Modern Times, 1350–1800* (New York, 2008), 174, see also infra 147–190, and Alan Debus, *The French Paracelsians: The Chemical Challenge to Medical and Scientific Tradition in Early Modern France* (Cambridge, UK, 2002).

tensions, paradoxes, and contradictions. Through shared work and the coordinated practices of men entangled with each other, with instruments, ships, and nature, a way could be navigated across these fluctuating frontiers. If observations were replicated across bodies, time, and media; if hands were steady, and instruments faithful; sailors and cosmographers could, given the proper motivation (e.g. a quest, a hunt, a mission), turn the tables on the world and translate distant spaces, currents, winds, and leaky ships into journeys across the world. Saints, poetry, astrolabes, sailors, and ancient historians; anatomies, hostages, and pepper, as well as scales, sails, miles of rope, logs, maps, histories, and bibles – all had their parts to play. At the same time, and at any point, resistance and contingent circumstances could intrude to block, alter, inflect, or reroute voyages, whether between pirates and trustworthy experts or ships moving from one end of the earth to the other. Extremes could never be entirely supplanted and stabilized; fictions could become facts, and facts fictions; treasonous acts might become sure translations, and misshapen monsters might become credible experts, or vice versa. Put another way, hubris might meet *Fortuna* at any moment. Despite refinements, fine tuning, and black boxing, encounters with nature could always be frustrated by the unexpected – a broken mast, a ripped sail, a diseased man, a contrary wind, an unseasonable storm, a flawed instrument, a faulty observation, or a mistaken theory.[14] Such contingencies could be disastrous, but they could also be generative of improvisation (if they didn't kill you first). Improvisations – experiments, hunches, and trials (*peira*) – as well as persistent errors, could thus find their way into repertoires of standard practice. But every new patch against the wild tremulous world leaking in could lead to new and unpredictable leaks. Equilibrium was difficult to achieve and hard to maintain. This, of course, was not only the case with wind and seas and balancing out, and appropriating – where possible – the forces of nature, but also in managing men (who were sick, tired, greedy, and hungry), and in interacting with (potentially hostile) strangers.

Success, by whatever means, whether by luck, chance or careful planning, carried with it its own kind of persuasion.[15] Confidence would grow with each success and with a proliferation of successors; that is, with replication and the movement not of one ship, but of many.[16] Trust, in

---

[14] Such as the magnetic declination of the compass as a solution to the problem of longitude.

[15] Even failures, such as the voyage of La Pensée and Le Sacre, were – or could be – successful if the circuit of a journey was completed, whether by ships or the (mis)information they carried.

[16] See Latour, "Visualization and Cognition."

this sense, was cumulative and applied not only to methods used, but to the human beings using them; indeed, both had to be synchronized to operate as an *escale* – a host of many acting as one. This routinization of replication through an accretion of voyages, skills and embedded instrumental knowledge can be seen in the translation of navigation on specific trade routes into standardized recipes, rules, tasks, and technologies. Put somewhat differently, the experience of sailors was entangled in the machinery of ships and the exigencies of nature; it was translated into charts, ropes, sails, instruments, tables and the routine choreographies of ritualized (black boxed) shipboard responsibilities; it was also converted into repertoires of (shipboard) command, dispute resolution and human relations, and into pigeons of cross-cultural communication and metrologies of trade. This kind of knowledge was not owned by one actor or another, but was rooted in an accumulation of intertwining observations, instrumental experiences, bodies of knowledge and body knowledges built up over time, voyage after voyage, generation after generation, in ships, instruments, and bodies. Out of sight of land, God-given reference points could still be found by those who knew how – in the motions of the sun, the moon, the stars and the mathematical rules that could be derived from them. Observed and noted, then captured on paper, wood, brass and iron, these abstract regularities could, in theory if not in practice, be translated to earth again and again and calibrated against men working capstans, cables, knots, and sails, and then compared to – and charted upon – calibrations embedded in previous iterations of the instruments, maps, and knowing bodies that provided a "first pass" on this data. Refinements, adjustments, discipline and repeated rehearsals were necessary, and necessarily ongoing.[17] Dispersed in practices, skills and knowledge distributed in work rituals, embodied in dispositions, ropes, canvas, and wood, inscribed in texts, instruments, and technologies, these human-material entanglements were given durability in both space and time – allowing them to be translated across different scales and in forms that could be adjusted, tweaked and replicated in response to changing circumstances and conditions. Through careful collection, inscription, and discipline, experience could be integrated into metrologies and matrices for scaling social, geographical, and epistemic distances. Thus could treason, counterfeits, fetishes, cunning, rhetoric, and monstrous men be transformed into credible (*bonne et loyale*) merchandise and trustworthy experts; though, as the sullied reputation of Parmentier and the legacy of the *Sea of Parmentiere* demonstrate, the

[17] Hutchins, *Cognition*, 96.

stability of such translations was always in danger of being traduced (see Chapter 6).

The expertise of new men such as Ango, Parmentier, and Crignon was empirical and they were empirics. The knowledge they possessed derived not from syllogistic deduction and philo-theological erudition, but from the entanglement of ideas, eyes, hands, and bodies in the world. How these knowledge practices were translated into social authority paralleled the extension of their cognitive authority into new domains of commerce and governmentality, but also, of course, into new domains of theology, that is, of speaking with, about, and for God. Becoming the embodied medium for the voice of nature was to claim unimpeachable social and intellectual authority. As we have seen, however, such claims relied on previously established hierarchies of status and power, on promises of profit and utility, on specific linguistic and cultural dispositions, on specific styles of thought and epistemic schemas, and, more often than not, on implied – and/or explicit – acts of violence, such as the shipboard dissections of members of La Pensée's crew, or Parmentier's execution of his Sumatran hostages.

The voyage of Le Sacre and La Pensée was a voyage through promiscuous couplings and uncouplings of "opposites" across generative tensions, momentary truces, and balances of power. It took place across vast geographical spaces, but also across social, epistemic, temporal, and spiritual divides. Terrestrial particularities were translated into poetic universals; menial mechanical tasks were given theological inflection; ancient authorities were brought back to life as trusted advisers; errors were logged as cartographic certainties; singular experiences were honed into embodied skills; observations of stars were translated into the running of courses; hostages and pepper were weighed and measured; sailors and ship-captains were transformed into emblematic (anti)heroes; and information was integrated into quests and, ultimately, institutionalized as disciplines. The map left by Parmentier's voyage traces out a history of constant and agonistic mediations, improvisational epiphanies, and ongoing and contested translations into ritual, rather than a definitive overcoming, elision, or purification: *Information* was intelligence and data; it was discipline and pilgrimage, it was to form, inform and transform; it was imperial, and it was a quest; *Expertise* was specialized and practiced knowledge; it was also a cunning, scheming and aspirational hunt for – and assertion of – social and epistemic authority; *Translation* was the rendering of one language into another; it was a physical displacement from one location to another, and it was the movement of knowledge and expertise from one discipline (and social status) to another; it was also conversion and co-optation, but it always bordered on treason; *Scale* was a disciplined host, it was the conquest of lands, seas and heavens by measure; it was a port-of-call, and the means by which men would be judged (weighed) in the life to come hereafter; *Confidence* was trust, and it was a con and a deception; it was credit, but with a variable interest rate; it was an assertion of unscalable divides and asymmetries of power, but it was also the working out of a balance and the reproduction of an equilibrium; *Replication*, at the end of it all, was to copy, to reply and to make a retort, but it was also to rehearse a recipe, to make a counterfeit, to prove and attest, to inscribe (record) and, of course, to induct future followers. Parmentier's voyage was a failed experiment – it was a piratical assay, a putting to proof, an attempt, and a trial – but it also launched manifold iterations that spanned great distances and breached (and in many senses, created) even greater divides. In charting the entangled assemblages of men, machines, poetry, nature, hostages, and scales, traced out by the journey of La Pensée and Le Sacre, we have then followed not only a journey across the world, from Dieppe to Sumatra and back again, but also a voyage into the (early) modern in the making.

Bibliography

Aiken, Jane Andrews. "Leon Battista Alberti's System of Human Proportions," *Journal of the Warburg and Courtauld Institutes* 43 (1980): 68–96.

Albano, Catrina. "Visible Bodies: Cartography and Anatomy," in A. Gordon and B. Klein (eds.), *Literature, Mapping, and the Politics of Space in Early Modern Britain* (Cambridge, 2001), 89–106.

Alexander, Amir. "The Imperialist Space of Elizabethan Mathematics," *Studies in History and Philosophy of Science* 26:4 (1995): 559–591.

Alter, Jean. *Les origines de la satire anti-bourgeoise en France: Moyen âge-XVIᵉ siècle* (Geneva, 1966).

Anderson, Benedict. *Imagined Communities* (London and New York, 1983).

Anonymous. "Jacobus Sylvius (Jacques Dubois) 1478–1555, Preceptor of Vesalius," *Journal of the American Medical Association* 195:13 (1966): 1147.

Anonymous. *Palinods, chants royaulx, ballades, rondeaulx et epigrammes à l'honneur de L'Immaculée Conception de la toute belle mère de Dieu Marie* (Paris and Caen, 1545).

Anonymous. *Actes de la Société. Procès-verbaux des séances, …* Presidence de M. Jomard, 19 décembre 1851, *Bulletin de la Société de Géographie* (July-December, 1951).

Arden, Heather. *Fools' Plays: A Study of Satire in the Sottie* (Cambridge, 1980).

Aristotle. *Posterior Analytics*, in J. Barnes (ed.), *The Complete Works of Aristotle: The Revised Oxford Translation* 2 vols (Princeton, 1984).

Ash, Eric H. *Power, Knowledge, and Expertise in Elizabethan England* (Baltimore, 2004).

Asseline, David. *Les antiquitez et chroniques de la ville de Dieppe* (Dieppe, 1874).

Badger, George P. (ed.), *The travels of Ludovico di Varthema in Egypt, Syria, Arabia Deserta and Arabia Felix, in Persia, India, and Ethiopia, A.D. 1503 to 1508*, introduced, trans. and ed. J. W. Jones (London, 1863).

Bakhtin, Mikhail. *Dialogic Imagination: Four Essays*, trans. and ed. Michael Holquist (Austin, TX and London, 1981).

Barckley, Richard. *Discourse of the Felicitie of Man* (London, 1598).

Baxter, James P. *A memoir of Jacques Cartier, sieur de Limoilou, his voyages to the St. Lawrence, a bibliography and a facsimile of the manuscript of 1534* (New York, 1906).

de Beauchamps, Pierre François Godard. *Recherches sur les théâtres de France* (Paris, 1735).

Beaujouan, Guy and E. Poulle. "Les origines de la navigation astronomique au XIVᵉ et XVᵉ siècles," M. Mollat and O. de Prat (eds.), *Le navire et l'économie maritime du XVᵉ aux XVIIIᵉ siècles* (Paris, 1957).

de Beaurepaire, Eugène de Robillard. *Les puys de palinod de Rouen et de Caen* (Caen, 1907).

Beck, Bernard. "Les Italiens de la mer. Marins et cartographes au service de la Normandie au XVIᵉ siècle," *Cahier des Annales de Normandie* 29 (2000). Les Italiens en Normandie, de l'étranger à l'immigré: Actes du colloque de Cerisy-la-Salle (8–11 octobre 1998): 129–142, 133–134.

Benedict, Philip. "French Cities from the Sixteenth Century to the Revolution: An Overview," in Philip Benedict (ed.), *Cities and Social Change in Early Modern France* (London and New York, 1992).

Rouen During the Wars of Religion (Cambridge, 1981).

Bennett, James A. *The Divided Circle: A History of Instruments for Astronomy, Navigation and Surveying* (Oxford, 1987).

"The Mechanics' Philosophy and the Mechanical Philosophy," *History of Science* 24 (1986): 1–28.

"The Travels and Trials of Mr. Harrison's Timekeeper," in Bourguet, Licoppe and Sibum (eds.), *Instruments, Travel and Science: Itineraries of Precision from the Seventeenth to the Twentieth Century* (London, 2000), 75–95.

Berce, Françoise. *Les premiers travaux de la commission des monuments historiques, 1837–1848* (Paris, 1979).

Bernard, Auguste. *Geofroy (sic) Tory, peintre et graveur, premier imprimeur royal, reformateur…* (Aubry, 1857).

Bernheimer, Richard. *Wild Men in the Middle Ages: A Study in Art, Sentiment, and Demonology* (Cambridge, MA, 1952).

Best, George. *A true discourse of the three Voyages of discouerie, for the finding of a passage to Cathaya, by the Northwest, vnder the conduct of Martin Frobisher Generall* in Richard Hakluyt and C. R. Beazley (eds.), *Voyages of Hawkins, Frobisher and Drake: Select Narratives from the "Principal Navigations" of Hakluyt* (Google eBook).

Biagioli, Mario. "The Social Status of Italian Mathematicians, 1450–1600," *History of Science* 27 (1989): 41–95.

Bibliothèque Municipale de Rouen, Ms. 1063 (Y. 16).

Bibliothèque nationale de France, Ms. Fr. 1537.

Bibliothèque nationale de France, Ms. Fr. 1739.

Bibliothèque nationale de France, Ms. Fr. 2205.

Bibliothèque nationale de France, Ms. Fr. 379.

Bibliothèque nationale de France, Ms. Fr. 594.

Bibliothèque nationale de France, na française 7510.

Blagrave, John. *The Mathematical Ievvel: shewing the making, and most excellent vse of a singuler instrument so called…* (London, 1585).

Blake, John. *West Africa: Quest for God and Gold* (London, 1977).

"New Light on Diogo Homem, Portuguese Cartographer," *The Mariner's Mirror* 28:2 (1942): 148–160.

Blix, Goran. *From Paris to Pompeii: French Romanticism and the Cultural Politics of Archaeology* (Philadelphia, 2013).

Blundeville, Thomas. *Exercises, containing six Treatises …* (London, 1594).

Bois, Guy. *The Crisis of Feudalism: Economy and Society in Eastern Normandy c. 1300–1550* (Cambridge, 1994).

Boiteux, L. A. *La fortune de mer. Le besoin de securité et les débuts de l'assurance maritime* (Paris, 1968).

Borges, J. L. *Collected Fictions*, trans. A. Hurley (New York, 1998).

Bouché-Leclercq, Auguste. *Notice sur la vie et les travaux de M. Charles Schefer* (Paris, 1899).

Bourdieu, Pierre. "Social Space and Symbolic Power," *Sociological Theory* 7:1 (1989): 14–25.

"Social Space and the Genesis of Groups," *Theory and Society* 14:6 (1985): 723–744.

Distinction: A Social Critique of the Judgement of Taste, trans. Richard Nice (Cambridge, MA, 1984).

"Le langage autorisé: Note sur les conditions sociales de l'efficacité du discours ritual," *Actes de la recherche en sciences sociales* 5:6 (1975): 183–190.

de Bovelles, Charles. *Sur les langues vulgaires et la variété de la langue française. Liber de differentia vulgarium linguarum et Gallici sermonis varietate … *trans. Colette Dumont-Demaizière (Paris, 1973).

Boyd, Barbara Weiden. "*Virtus Effeminata* and Sallust's Sempronia," *Transactions of the American Philological Association* 117 (1987): 183–201.

Brant, Sebastian. *La grant nef des folz du monde*, trans. Jean Drouyn (Lyon, 1499).

Bravo, Michael. "Ethnographic Navigation and the Geographical Gift," in David N. Livingstone and Charles W. J. Withers (eds.), *Geography and Enlightenment* (Chicago, 1999), 199–235.

Brotton, Jerry. *Trading Territories: Mapping the Early Modern World* (London, 1997).

Brown, Elizabeth A. R. "The Ceremonial of Royal Succession in Capetian France: The Funeral of Philip V," *Speculum* 55:2 (1980): 266–293.

Brown, Peter. *Society and the Holy in Late Antiquity* (Berkeley and Los Angeles, 1988).

The Cult of the Saints: Its Rise and Function in Latin Christianity (Chicago, 1981).

Brown, Robert. "Livy's Sabine Women and the Ideal of Concordia," in *Transactions of the American Philological Association* 125 (1995): 291–319.

Brunelle, Gayle. "Dangerous Liaisons: Mésalliance and Early Modern French Noble Women," *French Historical Studies* 19:1 (1995): 75–103.

The New World Merchants of Rouen, 1559–1630 (Kirksville, MO, 1991).

"Narrowing Horizons: Commerce and Derogation in Normandy," in Mack Holt (ed.), *Society and Institutions in Early Modern France* (Athens, GA, 1991), 63–79.

Bryant, Lawrence. *The King and the City in the Parisian Royal Entry Ceremony: Politics, Ritual, and Art in the Renaissance* (Geneva, 1986).

Buchet, Christian and Michel Vergé-Franceschi. *La mer, la France et l'Amérique latine* (Paris, 2006).

Butler, Travis. "Empeiria in Aristotle," *The Southern Journal of Philosophy* 41:3 (2003), 329–350.

Callon, Michel. "Some Elements of a Sociology of Translation: Domestification of the Scallops and Fishermen of St. Brieuc Bay," in J. Law (ed.), *Power,*

Action, Belief: A New Sociology of Knowledge? Sociological Review Monograph 32 (London, 1986).

Campbell, Ella. "Discovery and the Technical Setting: 1420–1520," *Terrae Incognitae* 8:1 (1976): 11–18.

Campbell, Mary Blaine. *Wonder and Science: Imagining Worlds in Early Modern Europe* (Ithaca, 1999).

Carlino, A. *Books of the Body. Anatomical Ritual and Renaissance Learning*, trans. J. Teddeschi and A. Tedeschi (Chicago, 1999).

Carroll, Stuart. *Blood and Violence in Early Modern France* (Oxford, 2006).

Carruthers, Mary. *The Book of Memory. A Study of Memory in Medieval Culture* (Cambridge, UK, 1990).

Castillo, Bernal Díaz del. *The History of the Conquest of New Spain*, edited and introduced by David Carrasco (Albuquerque, 2008).

Cave, Terence. *The Cornucopian Text: Problems of Writing in the French Renaissance* (Oxford, 1979).

Céard, Jean. *La nature et les prodiges: L'insolite au XVIe siècle* (Geneva, 1996).

Cesifo, Sheilah Ogilvie. "The Use And Abuse Of Trust: Social Capital And Its Deployment by Early Modern Guilds," Working Paper No. 1302 Category 10: Empirical And Theoretical Methods (October 2004).

Chabás, José and Bernard R. Goldstein. *Astronomy in the Iberian Peninsula: Abraham Zacut and the Transition from Manuscript to Print, Transactions of the American Philosophical Society*, New Series, 90:2 (2000).

Chaucer, Geoffrey. *A Treatise on The Astrolabe: addressed to his son Lowys*, ed. W. Skeat (London, 1872),

Clanchy, M. T. "Does Writing Construct the State?" *Journal of Historical Sociology* 15:1 (2002): 68–70.

From Memory to Written Record: England, 1066–1307 (Oxford, 1993).

Clark, Stuart. *Thinking with Demons: The Idea of Witchcraft in Early Modern Europe* (Oxford, 1999).

La Clavière, René de Maulde. *Jean Perréal dit Jean de Paris: peintre de Charles VIII, de Louis XII et de François Ier* (Paris, 1898).

Cohen, Adam Max. "Tudor Technology in Transition," in Kent Cartwright (ed.), *A Companion to Tudor Literature* (Sussex, 2010), 95–110.

Cohen, Kathleen. *Metamorphosis of a Death Symbol: The Transi Tomb in the Late Middle Ages* (Berkeley and Los Angeles, 1973).

Cohen, Paul. "Mediating Linguistic Difference in the Early Modern French Atlantic World: Linguistic Diversity in Old and New France," Working Paper in International Seminar on the History of the Atlantic World 1500–1800, Harvard University (2003).

Colin, Susi. "The Wild Man and the Indian in Early 16th Century Book Illustration," in C. Feest (ed.), *Indians and Europe: An Interdisciplinary Collection of Essays* (Aachen, 1987), 5–36.

Collins, Harry and R. Evans, *Rethinking Expertise* (Chicago, 2008).

Collins, Harry. *Changing Order: Replication and Induction in Scientific Practice* (Chicago, 1985).

"'Son of Seven Sexes', The Social Destruction of a Physical Phenomenon," *Social Studies of Science* 11 (1981): 33–62.

Conley, Tom. *The Self-Made Map: Cartographic Writing in Early Modern France* (Minneapolis and London, 1996).

Cook, Harold J. "Victories for Empiricism, Failures for Theory: Medicine and Science in the Seventeenth Century," in Charles T. Wolfe and Ofer Gal (eds.) *The Body as Object and Instrument of Knowledge: Embodied Empiricism in Early Modern Science* (Dordrecht, 2010), 9–32.

 Matters of Exchange: Commerce, Medicine, and Science in the Dutch Golden Age (New Haven, 2007).

 "The New Philosophy in the Low Countries," in R. Porter and M. Teich (eds.), *The Scientific Revolution in National Context* (Cambridge, 1992).

Cortés, Hernán. *Letters from Mexico*, trans., ed. and introduced by Anthony Pagden (New Haven and London, 1971).

Cortesão, Jaime. "The Pre-Columbian Discovery of America," *Geographical Journal* 89:1 (1937): 31–32.

Cotgrave, Randle. *Dictionarie of the French and English Tongues* (London, 1611).

Crignon, Pierre. *Poète et navigateur. Oeuvres en prose et en vers*, ed. John Nothnagle (Birmingham, AL, 1990).

Crosland, Jessie. "Prou, Preux, Preux Hom, Preud'ome," *French Studies* 1:2 (1947): 149–156.

Cummings, Anthony. *The Maecenas and the Madrigalist: Patrons, Patronage, and the Origins of the Italian Madrigal* (Philadelphia, 2004).

Curtius, Ernst Robert. *European Literature and the Latin Middle Ages* (New York, 1953).

Cuttler, Charles D. "Bosch and the *Narrenschiff*: A Problem in Relationships," *The Art Bulletin* 51:3 (1969): 272–276.

de Dainville, François. "How Did Oronce Fine Draw His Large Map of France?" *Imago Mundi*, 24 (1970): 49–55.

Daston, Lorraine and K. Park, *Wonders and the Order of Nature: 1150–1750* (New York, 1998).

Davies, Surekha. *Renaissance Ethnography and the Invention of the Human: New Worlds, Maps and Monsters* (Cambridge, 2016).

 "America and Amerindians in Sebastian Münster's Cosmographiae universalis libri VI (1550)," *Renaissance Studies* 25:3 (2011): 351–373.

Davis, Natalie Zemon. "Sixteenth-century Arithmetics on the Business Life," *Journal of the History of Ideas* 21 (1960): 18–48.

Day, Alan. *Historical Dictionary of the Discovery and Exploration of the Northwest Passage* (Lanham, Maryland, 2006).

Dear, Peter. "Mysteries of State, Mysteries of Nature: Authority, Knowledge and Expertise in the Seventeenth Century," in S. Jasanoff (ed.), *States of Knowledge: The Co-production of Science and Social Order* (London and New York, 2004), 206–224.

 Discipline and Experience: The Mathematical Way in the Scientific Revolution (Chicago, 1995).

Debus, Alan. *The French Paracelsians: The Chemical Challenge to Medical and Scientific Tradition in Early Modern France* (Cambridge, UK, 2002).

Defourneaux, Marcelin. *Les Français en Espagne aux XIᵉ et XIIᵉ siècles* (Paris, 1949).

Dermineur, Elise M. "The Civil Judicial System in France," *Frühneuzeit-Info* 23 (2012): 45–52.

Descola, Philippe. *Beyond Nature and Culture*, trans. Janet Lloyd (Chicago 2013).

Desmarquets, Jean-Antoine-Samson. *Mémoires chronologiques pour servir à l'histoire de Dieppe...* 2 Vols (Dieppe, 1785).

Desmont, M. "Le Port de Rouen et son commerce avec l'Amérique," *Société Normand de géographie* 33 (1911): 404–10.

Dewald, Jonathan. *Lost Worlds: The Emergence of French Social History, 1815–1970* (Philadelphia, 2010).

The Formation of a Provincial Nobility: The Magistrates of the Parlement of Rouen, 1499–1610 (Princeton, 1980).

Dictionnaire de L'Académie française (Paris, 1694).

Diffie, Bailey W. and G. D. Winius. *Foundations of the Portuguese Empire, 1415–1580* (Minneapolis, 1971).

Drakard, Jane. *A Kingdom of Words: Language and Power in Sumatra* (Oxford, 1999).

A Malay Frontier: Unity and Duality in a Sumatran Kingdom (Ithaca, 1990).

Dravasa, Etienne. *"Vivre noblement": Recherches sur la dérogeance de noblesse du XIVe au XVIe siècles* (Bordeaux, 1965).

Du Bellay, Joachim. *La defense et illustration de la language Française* (1549 edition), presented by Léon Séché (Paris, 1904).

La deffence, et illustration de la langue Françoyse. Ed. Jean-Charles Monferran (Geneva, 2001).

Dubois (Sylvius), Jacques. *Introduction à la langue Française suivie d'une grammaire* (1531), trans. and notes, C. Demaizière (Paris, 1998).

Eamon, William. *Science and the Secrets of Nature: Books of Secrets in Medieval and Early Modern Europe* (Princeton, 1994).

Edgerton, Samuel. *The Renaissance Rediscovery of Linear Perspective* (New York, 1975).

Elias, Norbert. *The Civilizing Process.* 2 vols. (New York, 1978).

Erasmus, Desiderius. *Collected Works of Erasmus, Colloquies*, Vol. 1, trans. Craig Thompson (Toronto, 1997).

Estancelin, Louis. *Études sur l'état actuel de la marine et des colonies françaises* (Paris, 1849).

Estienne, Robert. *Dictionarium latinogallicum* (Paris, 1552).

Estienne, Charles. *Tres libri de disectione partium corporis human* (Paris, 1545).

Fabri, Pierre. *Le grant et vray art de pleine rethorique...* (Rouen, 1534); also published in facsimile with an introduction by A. Héron, 2 vols (Rouen, 1890).

Farr, James. *The Work of France: Labor and Culture in Early Modern Times, 1350–1800* (New York, 2008).

Favre, Pierre Étienne Lazare. *Dictionnaire malais-français* (Paris, 1875).

Feest, Christian. "The People of Calicut: Objects, Texts, and Images in the Age of Proto-Ethnography," *Bol. Mus. Para. Emílio Goeldi. Cienc. Hum., Belém* 9:2 (2014): 287–303.

Ferguson, George. *Signs and Symbols in Christian Art* (Oxford, 1955).

Fernandez-Armesto, Felipe. *Amerigo: The Man Who Gave His Name to America* (New York, 2008).

Fernel, Jean. *Jean Fernel's The Hidden Causes of Things*, trans. John M. Forrester with introduction and annotations by J. Forrester and J. Henry (Leiden and Boston, 2005).

The Physiologia of Jean Fernel (1567), trans. John M. Forrester (Philadelphia, 2003).

Fiering, Norman. *The Jews and the Expansion of Europe to the West, 1450 to 1800* (New York and Oxford, 2001).

Fine, Oronce. *Liberalium disciplinarum Professoris Regii, Protomathesis …* (Paris, 1532).

Fisch, Max. "Vesalius and His Book," *Bulletin of the Medical Library Association* 8 (1943): 208–221.

Flint, Valerie. *The Imaginative Landscape of Christopher Columbus* (Princeton, 1992).

Focard, Jacques. *Paraphrase de l'astrolabe contenant: Les Principes de la geometrie, la sphere, l'astrolabe, ou déclaracion des parties de la terre* (Lyon, 1544).

da Fonseca, Luís Adão. "The Portuguese Military Orders and the Oceanic Navigations: From Piracy to Empire (Fifteenth to Early Sixteenth Centuries)," in M. Barber and J. M. Upton-Ward (eds.), *The Military Orders: On Land and By Sea* (Hampshire, 2008), 63–76.

Foucault, Michel. *Discipline and Punish: The Birth of the Prison* (New York, 1997).

"Of Other Spaces: Utopias and Heterotopias," in Neil Leach (ed.), *Rethinking Architecture: A Reader in Cultural Theory* (New York, 1997), 330–336.

"Politics and the Study of Discourse," in Graham Burchell, Colin Gordon, and Peter Miller (eds.), *The Foucault Effect: Studies in Governmentality* (Chicago, 1991), 53–72.

French, Roger K. "Natural Philosophy and Anatomy," in J. Ceard, M-M. Fontaine, and J-C. Margolin, *Le corps à la Renaissance. Actes du XXXe colloque de Tours, 1987* (Paris, 1990), 447–460.

Fries, Laurent. *Expositio vsusque astrolabij* (Strasbourg, 1522).

Frisch, Andrea. *The Invention of the Eyewitness: Witnessing and Testimony in Early Modern France* (Chapel Hill, 2004).

Froissart, *Chroniques*, ed. G. Raynaud (Paris, 1897).

Funkenstein, Amos. *Theology and the Scientific Imagination from the Middle Ages to the Seventeenth Century* (Princeton, 1986).

Fury, Cheryl. *Tides in the Affairs of Men: The Social History of Elizabethan Seamen, 1580–1603* (Westport, CT, 2001).

"The Work of G. V. Scammell," in C. Fury (ed.), *The Social History of English Seamen, 1485–1649* (Woodbridge, UK, 2012), 27–46.

Gadoffre, Gilbert. *La révolution culturelle dans la France des humanists* (Geneva, 1997).

de Gaetano, Armand L. "The Florentine Academy and the Advancement of Learning Through the Vernacular: the Orti Oricellari and the Sacra Accademia," *Bibliothèque d'Humanisme et Renaissance* 30:1 (1968): 19–52.

Galilei, Galileo. *Discoveries and Opinions of Galileo*, trans. and ed. by S. Drake (New York, 1957).

Gaffarel, Paul. "Anciens voyages Normands au Brésil," *Bulletin de la Société de l'histoire de Normandie* 5 (1887–1890): 236–239.

Jean Ango (Rouen, 1889).

Les colonies françaises (Paris, 1880).

"Obock," *Société normande de Géographie* (1879): 123–128.

Histoire du Brésil français au seizième siècle (Paris, 1878).

Gaude-Ferragu, Murielle. "Tombeaux et funérailles de coeur en France à la fin du moyen âge," in *Il cuore/The Heart* (Forence, 2003).

Gilbert, Felix. "Bernardo Rucellai and the Orti Oricellari: A Study on the Origin of Modern Political Thought," *Journal of the Warburg and Courtauld Institutes* 12 (1949): 101–131.

Godefroy, Théodore. *Le ceremonial François* (Cramoisy, 1649).

Gomez-Aranda, Mariano. "The Contribution of the Jews of Spain to the Transmission of Science in the Middle Ages," *European Review* 16:2 (2008): 169–181.

Goodrick-Clarke, Nicholas. *The Western Esoteric Traditions: A Historical Introduction* (Oxford, 2008).

Gosselin, Édouard and Charles de Beaurepaire, *Documents authentiques et inédits pour servir à l'histoire de la marine normande et du commerce rouennais pendant les XVIᵉ et XVIIᵉ siècles* (Rouen, 1876).

Goujard, Philippe. *La Normandie aux XVIᵉ et XVIIᵉ siècles: face à l'absolutisme* (Rennes, 2002).

Grandidier, A., J. Charles-Roux, Cl. Delhorbe, H. Froidevaux, and G. Grandidier (eds.), *Collection des ouvrages anciens concernant Madagascar*, 8 tomes (Paris, 1903), t. 1.

Gravier, Gabriel. *Jean Ango: Vicomte de Dieppe* (Rouen, 1903).

Les Normands sur la route des Indes (Rouen, 1880).

Greenblatt, Stephen. *Marvelous Possessions: The Wonder of the New World* (Chicago, 2008).

Gros, Gérard. *Le poète, la Vièrge et le prince du Puy: Étude sur la poésie mariale en milieu de cour aux XIVᵉ et XVᵉ siècles* (Paris, 1992).

Guenée, Bernard and Françoise Lehoux (eds.). *Les entrées royales françaises de 1328 à 1515* (Paris, 1968).

Guénin, Eugène. *La route de l'Inde* (Paris, 1903).

Ango et ses pilotes (Paris, 1901).

Histoire de la colonisation française (Paris, 1898).

Guérin, Léon. *Les navigateurs français: histoire des navigations, découvertes et colonisations françaises* (Paris, 1846).

Guéry, Abbé Ch. *Palinods ou puys de poésie en Normandie* (Evreux, 1916).

Gunderson, Erik. "The History of Mind and the Philosophy of History in Sallust's *Bellum Catilinae*," *Ramus: Critical Studies in Greek and Roman Literature* 29:2 (2000): 85–126.

Guyotjeannin, Olivier and Yann Potin. "La fabrique de la perpétuité le Trésor des Chartes et les archives du royaume (xiiiᵉ-xixᵉ siècle)," in *Revue de synthèse* (2004): 15–44.

Häberlein, Mark. *The Fuggers of Augsburg: Pursuing Wealth and Honor in Renaissance Germany* (Charlottesville, 2012).

Hampton, Timothy. *Literature and Nation in the Sixteenth Century: Inventing Renaissance France* (Ithaca, NY, 2001).

Hardy, Gaston Le. *De l'histoire du protestantisme en Normandie depuis son origine jusqu'à la publication de l'Édicte de Nantes* (Caen, 1869).

Harkness, Deborah. "Strange Ideas and English Knowledge," in Pamela Smith, Paula Findlen (eds.), *Merchants and Marvels: Commerce, Science, and Art in Early Modern Europe* (New York and London, 2013), 137–162.

Harrisse, Henry. *Americus Vespuccius: A Critical and Documentary Review of Two Recent English Books Concerning that Navigator* (London, 1895).

Head, Randolph. "Knowing Like a State: the Transformation of Political Knowledge in Swiss archives, 1470–1770," *Journal of Modern History* 75:4 (2003): 745–782.

Heller, Henry. *Labour, Science and Technology in France, 1500–1620* (Cambridge, 1996).

Heller-Roazen, Daniel. *The Enemy of All: Piracy and the Law of Nations* (New York, 2009).

Henry, John. "*Mathematics Made No Contribution to the Public Weal*: Why Jean Fernel (1497–1558) Became a Physician," *Centaurus* 53:3 (2011): 193–220.

Herwaarden, Jan van. "The Origins of the Cult of St James of Compostela," *Journal of Medieval History* 6:1 (1980): 1–35.

Higman, Francis. *Censorship and the Sorbonne: A Bibliographical Study of Books in French Censured by the Faculty of Theology of the University of Paris, 1520–1551* (Geneva, 1979).

Hill, Christine. "Symphorien Champier's Views on Education in the *Nef des princes* and the *Nef des dames vertueuses*," *French Studies* 7:4 (1953): 323–334.

Hill, Christopher. *National History and the World of Nations: Capital, State, and the Rhetoric of History in Japan, France, and the United States* (Durham, 2008).

Hoffman, B. G. "Account of a Voyage Conducted in 1529 to the New World, Africa, Madagascar, and Sumatra, Translated from the Italian, with Notes and Comments," *Ethnohistory* 10:1 (1963), 1–31, 33–79.

Hofstadter, Richard. *Anti-intellectualism in American Life* (New York, 1962).

Howse, Derek. "Navigation and Astronomy," *Renaissance and Modern Studies* 30 (1986): 62–3.

Greenwich Time and the Discovery of Longitude (Oxford, 1980).

Huë, Denis. *La poésie palinodique à Rouen: 1486–1550* (Paris, 2002).

Petite anthologie palinodique: 1486–1550 (Paris, 2002).

"Un nouveau manuscrit palinodique, Carpentras, Bibliothèque Inguimbertine no. 385," *Le Moyen Français* 35–36 (1995): 175–230.

Huppert, George. *Les Bourgeois Gentilshommes: An Essay on the Definition of Elites in Renaissance France* (Chicago, 1977).

Hutchins, Edwin. *Cognition in the Wild* (Cambridge, MA, 1995).

Ivins, William. "Geoffroy Tory," *The Metropolitan Museum of Art Bulletin*, 15:4 (1920): 79–86.

Jacquinot, Dominique. *L'usaige de l'astrolabe* (Paris, 1545).

Jager, Eric. *Book of the Heart* (Chicago, 2000).

Jenn, Jean-Marie, Françoise Jenn, Jean-Pierre Babylon, and Alain Erlande-Bradenbourg (eds.). "Le roi, la sculpture et la mort: Gisants et tombeaux de la Basiliqe de Saint Denis," *Archives departmentales de la Seine-Saint-Denis, Bulletin* 3 (June, 1975).

Johnson, Christine. *The German Discovery of the World: Renaissance Encounters with the Strange and Marvelous* (Charlottesville and London, 2008).

Joukovsky, Françoise. *La gloire dans la poésie française au XVIᵉ siècle* (Geneva, 1969).

Julien, Charles-André. *Les débuts de l'expansion et de la colonization françaises* (Paris, 1947).

Kantorowicz, Ernst. *The King's Two Bodies: A Study in Medieval Political Theology* (Princeton, 1957).

"The 'King's Advent' and the Enigmatic Panels in the Doors of Santa Sabina," *Art Bulletin* 26:4 (December, 1944): 207–231.

Kaplow, Lauren. "Redefining *Imagines*: Ancestor Masks and Political Legitimacy in the Rhetoric of New Men," *Mouseion* 3:8 (2008): 409–416.

Karrow, Robert W. *Mapmakers of the Sixteenth Century and Their Maps: Bio-Bibliographies of the Cartographers of Abraham Ortelius, 1570* (Chicago, 1993).

Kaysersberg, Johannes Geiler von. *Navicula penitentie* (Strasburg, 1512).

Kellenbenz, H. "Autour de 1600: le commerce du poivre des Fugger et le marché international du poivre," *Annales. Économies, Sociétés, Civilisations* 11:1 (1956): 1–28.

Kellett, Charles E. "Sylvius and The Reform of Anatomy," *Medical History* 5:2 (1961).

Kellym, Gordon P. *A History of Exile in the Roman republic* (Cambridge and New York, 2006).

Kempe, Michael. "'Even in the Remotest Corners of the World': Globalized Piracy and International Law, 1500–1900," *Journal of Global History* 5:3 (2010): 353–72.

Kosto, Adam. *Hostages in the Middle Ages* (Oxford, 2012).

Kristeller, Paul Oskar. "The Modern System of the Arts," in P. Kristeller, *Renaissance Thought II: Papers on Humanism and the Arts* (New York, 1965), 163–227.

Kristeva, Julia. *Language – The Unknown: An Initiation Into Linguistics* (New York, 1989).

Kuhn, Thomas. *The Essential Tension: Selected Studies in Scientific Tradition and Change* (Chicago and London, 1977).

Kumin, Beata. "Useful to Have, but Difficult to Govern. Inns and Taverns in Early Modern Bern and Vaud," *Journal of Early Modern History* 3:2 (1999).

Laborde, Léon de. *La Renaissance des arts a la cour de France*, t. I (Paris, 1853).

Laborie, Jean-Claude and Frank Lestringant (eds.). *Histoire d'André Thevet Angoumoisin, cosmographe du roy, de deux voyages par luy faits aux Indes australes, et occidentales* (Geneva, 2006).

Lach, Donald F. *Asia in the Making of Europe*, Volume I: *The Century of Discovery*, Book 1 (Chicago, 1994).

Ladurie, Emmanuel Le Roy. *The French Peasantry, 1450–1660*, trans. A. Sheridan (Berkeley and Los Angeles, 1987).

Latour, Bruno. *We Have Never Been Modern* (Cambridge, MA, 1993).

Science in Action: How to Follow Scientists and Engineers through Society (Cambridge, MA, 1987).

"Visualization and Cognition," *Knowledge and Society* 6 (1986): 1–40.

"Give me a Laboratory and I'll Raise the World," in Karin Knorr-Cetina and Michael Mulkay (eds.), *Science Observed: Perspectives on the Social Study of Science* (London, Beverly Hills and New Delhi, 1983), 141–170.

Law, John. "On the Methods of Long-distance Control: Vessels, Navigation and the Portuguese Route to India," in J. Law (ed.), *Power, Action and Belief: A New Sociology of Knowledge?* (London, 1986), 234–263.

Lebas, Georges. *Les Palinods et les poètes dieppois* (Dieppe, 1904).

Lebègue, Raymond. "Un chant royal d'humaniste," *Bibliothèque d'Humanisme et Renaissance* 18:3 (1956): 432–435.

Leblanc, Vincent. *Les voyages fameux du Sieur V. Leblanc* (1648).

Lecoq, Anne-Marie. *François Ier imaginaire: symbolique et politique à l'aube de la Renaissance française* (Paris, 1987).

Legris, Albert. *L'église Saint-Jacques de Dieppe: notice historique & descriptive* (Dieppe, 1918).

Leitch, Stephanie. *Mapping Ethnography in Early Modern Germany: New Worlds in Print Culture* (New York, 2010).

Levine, Timothy R., Rachel K. Kim, and Lauren M. Hamel. "People Lie for a Reason: Three Experiments Documenting the Principle of Veracity," *Communication Research Reports* 27:4 (2010).

Lingo, Alison Klairmont. "Empirics and Charlatans in Early Modern France: The Genesis of the Classification of the 'Other' in Medical Practice," *Journal of Social History* 19:4 (1986): 83–603.

Lobur, John Alexander. *Consensus, Concordia, and the Formation of Roman Imperial Ideology* (New York, 2008).

Long, Pamela. *Openness, Secrecy, Authorship: Technical Arts and the Culture of Knowledge from Antiquity to the Renaissance* (Baltimore, 2001).

Luchet, Auguste and Félix Pyat. *Ango: drame en cinq actes, six tableaux, avec un epilogue* (Paris, 1835).

Lynch, Michael. "Circumscribing Expertise: Membership Categories in Courtroom Testimony," in S. Jasanoff (ed.), *States of Knowledge: The Co-production of Science and Social Order* (London and New York, 2004), 161–80.

Lyons, John. *Exemplum* (Princeton, 1989).

McCormick, Michael. *Origins of the European Economy: Communications and Commerce AD 300–900, Parts 300–900* (Cambridge, 2001).

McGhee, R. *Arctic Voyages of Martin Frobisher: An Elizabethan Venture* (Montreal and London, 2001).

Machiavelli, Niccolò. *Art of War*, trans. and ed. C. Lynch (Chicago, 2009).

McKeon, Michael. *The Origins of the English Novel: 1600–1740* (Baltimore, 2002).

Madelaine, Victor. *Le Protestanisme dans le pays de Caux* (Paris, 1906).

Maillard, Jean. BN Ms fr. 1382, *Description de tous les portz de mer de l'univers*.

Mâle, Émile. *L'art religieux du XIIe siècle en France: étude sur les origines de l'iconographie du moyen âge* (Paris, 1922).

Mangani, Giorgio. "Abraham Ortelius and the Hermetic Meaning of the Cordiform Projection," *Imago Mundi* 50 (1998): 59–83.

Margry, Pierre. *Les navigations françaises et la révolution maritime du XIVe au XVIe siècle* (Paris, 1867).

Familles de la France colonial (Paris, 1851).

Marsden, William. *The History of Sumatra, Containing an Account of the Government, Laws, Customs and Manners of the Native Inhabitants with a Description of the Natural Productions, and a Relation of the Ancient Political State of that Island* (London, 1811).

Mason, Peter. *The Lives of Images* (London, 2001).

Masse, Vincent. "Les 'sept hommes sauvages' de 1509: Fortune éditoriale de la première séquelle imprimée des contacts franco-amérindiens," in Andreas Motsch & Grégoire Holtz (eds.), *Éditer la Nouvelle-France* (Laval, 2011), 83–106.

Massing, Jean Michel. "Hans Burgkmair's Depiction of Native Africans," *RES: Anthropology and Aesthetics* 27 (1995): 39–51.

"Early European Images of America: the Ethnographic Approach," in Jay A. Levenson (ed.), *Circa 1492: Art in the Age of Exploration* (New Haven, 1991).

Mauelshagen, Franz. "Networks of Trust: Scholarly Correspondence and Scientific Exchange in Early Modern Europe," *The Medieval History Journal* 6:1 (2003): 1–32.

Mauss, Marcel. *The Gift: Forms and Functions of Exchange in Archaic Societies*, trans. I. Cunnison (New York and London, 1967).

Mercier, P. G. Le. *Histoire du théâtre François, depuis son origine jusqu'à present* (Paris, 1745).

Meyer, Barbara Hochstetler. "Jean Perréal and Portraits of Louis XII," *The Journal of the Walters Art Gallery* 40 (1982): 41–56.

Mialet, Hélène. *Hawking Incorporated: Stephen Hawking and the Anthropology of the Knowing Subject* (Chicago, 2012).

L'entreprise créatrice (Paris, 2008).

Mignolo, Walter. *The Darker Side of the Renaissance: Literacy, Territoriality and Colonization* (Ann Arbor, MI, 1995).

Miller, John D. *Beads and Prayers: The Rosary in History and Devotion* (London, 2002).

Minard, Philippe. *La fortune du colbertisme* (Paris, 1998).

Mittman, Asa Simon and Peter J. Dendle. *The Ashgate Research Companion to Monsters and the Monstrous* (Surrey, 2013).

Mollat, Michel. "Anciens voyages Normands au Brésil," *Bulletin de la Société de l'histoire de Normandie* 5 (Rouen, 1987 and 1990), 236–9 and 249–67.

Mollat, Michel. *Histoire de Rouen* (Toulouse, 1979).

Études d'histoire maritime: 1938–75 (Torino, 1977).

Le commerce maritime normand á la fin du Moyen Age: Étude d'histoire économique et sociale (Paris, 1952).

Mollat, Michel and Jacques Habert. *Giovanni et Girolamo Verrazano, navigateurs de François I*ᵉʳ (Paris, 1982).

Momigliano, Arnaldo. "Ancient History and the Antiquarian," *Journal of the Warburg and Courtauld Institutes* 13:3/4 (1950): 285–315.

Montaiglon, Anatole de. "La sculpture française à la Renaissance: La Famille des Justes en France," *Gazette des Beaux-Arts* 12 (November, 1875).

Montaigne, Michel de. *The Complete Essays*, trans. D. Frame (Stanford, 1965).

Moore, Jr., John K. "Santiago's Sinister Hand: Hybrid Identity in the Statue of Saint James the Greater at Santa Marta de Tera," *Peregrinations: Journal of Medieval Art & Architecture* 4:3 (2014): 31–62.

"Juxtaposing James the Greater: Interpreting the Interstices of Santiago as Peregrino and Matamoros," *La Corónica: A Journal of Medieval Hispanic Languages, Literatures, and Cultures* 36:2 (2008): 313–344.

Moore, Robert I. *The War on Heresy* (Cambridge, MA, 2012).

Murphy, Agnes. *The Ideology of French Imperialism, 1871–1881* (New York, 1948).

Murphy, Kevin D. *Memory and Modernity: Viollet-le-Duc at Vézelay* (Philadelphia, 1999).

Murray, James. "Of Nodes and Networks: Bruges and the Infrastructure of Trade in the Fourteenth-Century," in Peter Stabel, Bruno Blondé, and Anke Greve (eds.), *International Trade in the Low Countries, 14th–16th Centuries* (Leuven 2000).

Navare, Marguerite De. *Lettres de Marguerite d'Angoulême, soeur de François Ier, Reine de Navare*, ed. F. Génin (Paris, 1841).

Neal, Katherine. "The Rhetoric of Utility: Avoiding Occult Associations for Mathematics through Profitability and Pleasure," *History of Science* 37 (1999): 151–178.

Nelles, Paul. "Renaissance Libraries," D. Stam (ed.), *International Dictionary of Library History* (2001), 134–151.

Newcomer, Charles B.. "The Puy at Rouen," *Publications of the Modern Language Association of America* 31 (1916): 211–231.

Nicholls, David. "Social Change and Early Protestantism in France: Normandy, 1520–62," *European Studies Review* 10:3 (1980): 279–308.

Nichols, Stephen G. *Romanesque Signs: Early Medieval Narrative and Iconography* (New Haven, 1983).

Nicot, Jean. *Thresor de la langue française* (Paris, 1606).

Nora, Pierre. "Between Memory and History: Les Lieux de Mémoire," *Representations* 26 (Spring, 1989): 7–24.

North, John D. "Essay Review: Some Jewish Contributions to Iberian Astronomy" *Aleph: Historical Studies in Science and Judaism* 2 (2002): 271–278.

O'Connor, Dorothy. "Notes on the Influence of Brant's 'Narrenschiff' outside Germany," *The Modern Language Review* 20:1 (1925): 64–70.

O'Connor, Edward D. (ed.). *The Dogma of the Immaculate Conception* (Notre Dame, 1958).

O'Malley, Charles D. *Vesalius of Brussels, 1514–1564* (Berkeley and Los Angeles, 1964).

Orth, Myra. "The Triumphs of Petrarch illuminated by Godefroy le Batave," *Gazette des beaux-arts* (1984), 197–206.

"Two Books of Hours for Jean Lallemant Le Jeune," *The Journal of the Walters Art Gallery* 38 (1980): 70–93.

Osmond, Patricia. "'Princeps Historiae Romanae': Sallust in Renaissance Political Thought," *Memoirs of the American Academy in Rome* 40 (1995).

Oursel, Charles. *Notes pour servir à l'histoire de la Réforme en Normandie au temps de François Ier* (Caen, 1913).

Panel, Gustave (ed.). *Documents concernant les pauvres de Rouen, 1224–1634* (Paris, 1917).

Panofsky, Erwin. *Perspective as Symbolic Form*, trans. C. Wood (New York, 1997).

Tomb Sculpture. Four Lectures on its Changing Aspects (New York, 1964).

Park, Katharine. "The Life of the Corpse: Division and Dissection in Late Medieval Europe," *Journal of the History of Medicine and Allied Sciences* 50:1 (1995): 111–132.

"The Criminal and the Saintly Body: Autopsy and Dissection in Renaissance Italy," *Renaissance Quarterly* 47:1 (1994): 1–33.

Parmentier, Jean. *Oeuvres poétiques*, ed. F. Ferrand (Geneva, 1971).

Pérez-Mallaína, Pablo E. *Spain's Men of the Sea: Daily Life on the Indies Fleets in the Sixteenth Century*, trans. C. R. Philips (Baltimore, 1998).

Péricatd-Méa, Denise. *Compostelle et cultes de Saint-Jacques au Moyen Âge* (Paris, 2000).

Perry, Horace. *The Age of Reconnaissance* (Los Angeles and Berkeley, 1981).

Petrarch. *Epistole familiari*, ed. V. Rossi. 4 vols (Florence, 1933–42).

Peutinger, Konrad. *Sermones conuiuales* (Strasburg, 1506).

Picot, Émile. *Notice sur Jacques Le Lieur echevin de Rouen sur ses Heures manuscrites* (Rouen, 1913).

Picot, Émile (ed.). *Recueil général des sotties*. 3 vols (Paris, 1902–12).

(ed.). *Théatre mystique de Pierre du Val et des libertins spirituels de Rouen, au XVIe siècle* (Paris, 1882).

Pietz, William. "The Problem of the Fetish –3a," *Res* 16 (1988): 105–123.

"The Problem of the Fetish –2," *Res* 13 (1987): 23–45.

"The Problem of the Fetish –1," *Res* 9 (1985): 5–17.

Pimentel, Juan. *Testigos del mondo: Ciencia literatora y viajes en la illustración* (Madrid, 2003).

Pinheiro, Cláudio Costa. "Words of Conquest: Portuguese Colonial Experiences and the Conquest of Epistemological Territories," *Indian Historical Review* 36:1 (2009): 37–53.

Pires, Tomé. *The Suma Oriental of Tome Pires. Account of the East, from the Red Sea to China, written in Malacca and India in 1512–1515*, Cortesao, Armando (ed.), 2 Volumes (New Delhi, 2005).

Plato's Republic. Trans. B. Jowett (Cambridge, MA, 2008).

Poblacion, Ioannis Martin. *De vsu astrolabi compendium* (Paris, 1500 and 1527).

Pomian, Krzysztof. "The Archives: From the *Trésor des Chartes* to the CARAN," in Pierre Nora (ed.), *Rethinking France: Les Lieux De Mémoire*, Volume 4: *Histories And Memories* (Chicago, 2001), 27–100.

Popper, Karl. *The Logic of Scientific Discovery* (London and New York, 1959).

Porter, Pamela. *Courtly Love in Medieval Manuscripts* (Toronto, 2003).

Portuondo, María. *Secret Science: Spanish Cosmography and the New World* (Chicago, 2009).

Pradel, Pierre. *Michel Colombe, le dernier imagier gothique* (Paris, 1953).

Prasad, Amit. *Imperial Technoscience: Transnational Histories of the MRI in the United States, Britain, and India* (Cambridge, MA, 2014).

Prentout, Henri. "La Reformation en Normandie et le debuts de la Réforme à l'université de Caen," *Revue Historique* 114 (1913): 285–305.

Pumfrey, Stephen. "Who Did the Work? Experimental Philosophers and Public Demonstrators in Augustan England," *British Journal for the History of Science* 28:2 (1995): 131–156.

Quintilianus, Marcus Fabius. *The Institutio Oratoria of Quintilian*, E. Capps, T. E. Page, and W. H. D. Rouse (eds.), trans. H. E. Butler. 4 vols (London, 1922).

Rabelais, François. *Five books of the lives, heroic deeds and sayings of Gargantua and his son Pantagruel*, trans. Sir Thomas Urquhart of Cromarty and Peter Antony Motteux (New York, 2005).

Oeuvres de François Rabelais contenant la vie de Gargantua et celle de Pantagruel (Paris, 1854).

Raj, Kapil. "Beyond Postcolonialism… and Postpositivism: Circulation and the Global History of Science," *Isis* 104:2 (2013): 337–347.

Relocating Modern Science: Circulation and the Construction of Knowledge in South Asia and Europe, 1650–1900 (Basingstoke and New York, 2007).

"Go-Betweens, Travelers, and Cultural Translators," in B. Lightman (ed.), *A Companion to the History of Science* (Chichester, 2016), 39–57.

Ramusio, Giovanni B. *Navigationi et Viaggi,* Vol. 3 (Venice, 1556).

Randles, W. G. L. *Portuguese and Spanish Attempts to Measure Longitude in the 16th Century* (Coimbra, 1984).

"La diffusion dans l'Europe du XVIᵉ siècle des connaissances géographique dues aux découvertes portugaises," in Randles, *Geography, Cartography and Nautical Science in the Renaissance* (Burlington and Hampshire, 2000), 269–277.

"The Emergence of Nautical Astronomy in Portugal in the XVth century," *Journal of Navigation* 51:1 (1998): 46–57.

Raulston, Stephen B. "The Harmony of Staff and Sword: How Medieval Thinkers Saw Santiago Peregrino and Matamoros," *La corónica: A Journal of Medieval Hispanic Languages, Literatures, and Cultures* 36:2 (2008): 345–367.

Reid, Anthony. *Charting the Shape of Early Modern Southeast Asia* (Singapore, 2000).

Southeast Asia in the Age of Commerce, 1450–1680: The Lands Below the Winds, Volume 1 (New Haven, 1988).

Reid, Dylan. "Patrons of Poetry: Rouen's Confraternity of the Immaculate Conception of Our Lady," in A. van Dixhoorn and S. S. Sutch (eds.), *The Reach of the Republic of Letters: Literary and Learned Societies in Late Medieval and Early Modern Europe*, 2 vols (Leiden, 2008), I: 33–78.

Reiss, Timothy. *Knowledge, Discovery and Imagination in Early Modern Europe: The Rise of Aesthetic Rationalism* (Cambridge, 1997).

Renouard, Philippe. *Imprimeurs parisiens, libraires, fondeurs de caractères et correcteurs d'imprimerie*, Google eBook (Paris, 1898).

Revel, Jacques. "The Uses of Civility" (167–205) in Philippe Ariès and Georges Duby (eds.), *A History of Private Life*, volume III, Roger Chartier (ed.), *Passions of the Renaissance* (Cambridge, MA, 1989).

Rice, Eugene. *The Prefatory Epistles of Jacques Lefèvre d'Étaples* (New York and London, 1972).

La Roncière, Charles de. *Histoire de la marine française*, 3 tomes (Paris, 1899–1906).

Rubies, Joan-Pau. *Travel and Ethnology in the Renaissance: South India through European Eyes, 1250–1625* (Cambridge, 2000).

Rubin, Miri. *Corpus Christi the Eucharist in Late Medieval Culture* (Cambridge, 1991).

Rucquoi, Adeline (ed.). *Saint-Jacques et la France: actes du colloque des 18 et 19 janvier 2001 à la fondation Singer-Polignac* (Paris, 2003).

Rummel, Erica. *The Humanist–Scholastic Debate in the Renaissance and Reformation* (Cambridge and London, 1995).

Ruysch, Johannes. *Universalior cogniti orbis tabula, ex recentibus confecta observationibus* (1508).

Sallust, Gaius Sallustius Crispus. *L'histoire catilinaire, composée par Saluste, hystorian romain, et translatée par forme d'interprétation d'ung très brief et élégant latin en nostre vulgaire françoys par Jehan Parmentier, marchant de la ville de Dieppe* (Paris, 1528).

The Histories, trans. J. C. Rolfe (Cambridge, MA, 1951).

Salmon, John Hearsey McMillan. "Storm over the Noblesse," *Journal of Modern History* 53:2 (1981): 242–57.

Sanders, P. M. "Charles de Bovelles's Treatise on the Regular Polyhedra (Paris, 1511)," *Annals of Science* 41:6 (1984): 513–66.

Sandman, Alison. "Controlling Knowledge: Navigation, Cartography, and Secrecy in the Early Modern Spanish Atlantic," in James Delbourgo and Nicholas Dew (eds.), *Science and Empire in the Atlantic World* (New York and London, 2008), 31–51.

"Mirroring the World: Sea Charts, Navigation, and Territorial Claims in Sixteenth-century Spain," in P. H. Smith and P. Findlen (eds.), *Merchants and Marvels: Commerce, Science, and Art in Early Modern Europe* (New York, 2002), 83–108.

Saunders, H. S. *All the Astrolabes* (Oxford, 1984).

Sawday, Jonathan. *The Body Emblazoned: Dissection and the Human Body in Renaissance Culture* (New York, 1995).

Schaffer, Simon. "Les cérémonies de la mesure: Repenser l'histoire mondiale des sciences," *Annales. Histoire, Sciences Sociales* 2 (2015): 409–435.

"Glass Works: Newton's Prisms and the Uses of Experiment," in D. Gooding, T. Pinch and S. Schaffer (eds.), *The Uses of Experiment* (Cambridge, 1989), 67–104.

Schaffer, Simon and S. Shapin. *Leviathan and the Air-Pump: Hobbes, Boyle, and the Experimental Life* (Princeton, 1985).

Schalk, Ellery. *From Valor to Pedigree: Ideas of Nobility in France in the Sixteenth and Seventeenth Centuries* (Princeton, 1986).

Schefer, Charles. *Le discours de la navigation de Jean et Raoul Parmentier de Dieppe: voyage à Sumatra en 1529: description de l'isle de Sainct-Dominigo* (Paris, 1883).

Schmitt, Christian. "Bovelles, Linguist," in Actes du Colloque international tenu à Noyon, *Charles de Bovelles en son cinquième centenaire: 1479–1979* (Paris, 1982), 247–263.

Schuller, Rudolph. "The Oldest Known Illustration of South American Indians," *Journal de la Société des Américanistes* 16:1 (1924): 111–118.

Selinger, Evan and R. Crease (eds.). *The Philosophy of Expertise* (New York, 2006).

Serres, Michel. *The Natural Contract* (Ann Arbor, 1992).

Shackleton Bailey, David R. "Nobiles and Novi Reconsidered," *The American Journal of Philology* 107:2 (1986): 255–260.

Shakespeare, William. *The Tragicall Historie of Hamlet Prince of Denmarke* (London, 1603).

Shapin, Steven. *The Social History of Truth: Civility and Science in Seventeenth-Century England* (Chicago, 1994).

Shapin, Stevin and S. Schaffer. *Leviathan and the Air-Pump: Hobbes, Boyle, and the Experimental Life* (Princeton, 1985).

Shell, Marc. *Art and Money* (Chicago, 1995).

Sherberg, Michael. "The Accademia Fiorentina and the Question of the Language: The Politics of Theory in Ducal Florence," *Renaissance Quarterly* 56:1 (2003): 26–55.

Sherrington, Charles. *The Endeavour of Jean Fernel* (Cambridge, 1946).

Sibum, Otto. "Reworking the Mechanical Value of Heat: Instruments of Precision and Gestures of Accuracy in Early Victorian England" in *Studies in history and philosophy of science* 26:1 (1995): 73–106.

Skrine, Peter. "The Destination of the Ship of Fools: Religious Allegory in Brant's 'Narrenschiff," *The Modern Language Review* 64:3 (1969): 576–96.

Smith, Pamela. *The Body of the Artisan: Art and Experience in the Scientific Revolution* (Chicago, 2004).

Smyser, H. M. (trans. and ed.). *The Pseudo-Turpin* (Cambridge, MA, 1937).

Sousa, Ronald. "Cannibal, Cartographer, Soldier, Spy: The Peirai of Mendes Pinto's *Peregrinação*," in Elizabeth Fowler and Roland Greene (eds.), *The Project of Prose in Early Modern Europe and the New World* (Cambridge, UK, 1997), 15–30.

Stimson, Alan. *The Mariner's Astrolabe: A Survey of Known Surviving Sea Astrolabes* (Utrecht, 1988).

Sturtevant, William. "First Visual Images of Native America," in Fredi Chiappelli, J. Michael, B. Allen, Robert Louis Benson (eds.), *First Images of America: The Impact of the New World on the Old*, 2 Vols (Berkeley and Los Angeles, 1976), I: 417–454.

Subrahmanyam, Sanjay. *Explorations in Connected History: From the Tagus to the Ganges* (Oxford, 2005).

"Connected Histories: Notes towards a Reconfiguration of Early Modern Eurasia," in *Beyond Binary Histories: Re-imagining Eurasia to c.1830*, ed. Victor Lieberman (Ann Arbor, 1999), 289–316.

The Career and Legend of Vasco Da Gama (Cambridge, 1998).

Suze, Zijlstra. "To Build and Sustain Trust: Long-Distance Correspondence of Dutch Seventeenth-Century Merchants," *Dutch Crossing: Journal of Low Countries Studies* 36:2 (2012): 114–131.

Syme, R. *Sallust* (Berkeley and Los Angeles, 1964).

Taylor, E. G. R. *Tudor Geography: 1485–1583* (London, 1930).

"Jean Rotz: His Neglected Treatise on Nautical Science," *The Geographical Journal* 73:5 (1929): 455–459.

Thompson, Janice. *Mercenaries, Pirates, and Sovereigns: State-Building and Extraterritorial Violence in Early Modern Europe* (Princeton, 1996).

Tilly, Arthur. *Studies in the French Renaissance* (Cambridge, 1922).

Toulouse, Sarah. "Marine Cartography and Navigation in Renaissance France," in David Woodward (ed.), *The History of Cartography*, Vol. 3 (Chicago, 2007), 1550–1568.

Tresch, John. *The Romantic Machine: Utopian Science and Technology after Napoleon* (Chicago, 2012).

Trivellato, Francesca. *The Familiarity of Strangers: The Sephardic Diaspora, Livorno, and Cross-Cultural Trade in the Early Modern Period* (New Haven and London, 2009).

Trueta, J. "The Contribution Of Michael Servetus to the Scientific Development of the Renaissance," *The British Medical Journal* 2 (1954): 507–510.

Turgeon, Laurier. "Codfish, Consumption, and Colonization: The Creation of the French Atlantic World During the Sixteenth Century," in Caroline A. Williams (ed.), *Bridging the Early Modern Atlantic World: People, Products, and Practices on the Move* (Surrey and Burlington, 2009), 33–56.

Turner, Anthony. *Early Scientific Instruments: Europe 1400–1800* (London, 1987).

Vanden Broecke, Steven. "The Use of Visual Media in Renaissance Cosmography: the Cosmography of Peter Apian and Gemma Frisius," *Paedagogica Historica*, 36:1 (2000): 131–150.

Van Duzer, Chet. *The World for a King: Pierre Desceliers' World Map of 1550* (London, 2015).

Vauchez, André. "The Cathedral," in Pierre Nora, Lawrence D. Kritzman (eds.), *Realms of Memory: Traditions* (New York, 1967).

Velho, Álvaro. *A Journal of the First Voyage of Vasco Da Gama, 1497–1499* (London, 1898).

Verdier, Pierre Le (ed.). *Pierre du Val, le puy du souverain amour*, originally published 1543 (Rouen, 1920).

Vesely, Dalibor. "The Architectonics of Embodiment" and Robert Tavernor, "Contemplating Perfection through Piero's Eyes," in George Dodds and Robert Tavernor (eds.), *Body and Building: Essays on the Changing Relation of Body and Architecture* (Cambridge, MA., 2002), 28–43 and 78–93.

Vidoue, Pierre. *Palinodz, chants royaulx, ballades, rondeaulx, et epigrammes, a lhonneur de l'immaculee Conception de la toute belle mere de dieu Marie Patronne de Normans presentez au puy a Rouen …* (sl, 1525). Reprinted by E. de Robillard de Beaurepaire (Rouen, 1897).

Vigarié, A. C. "France and the Great Maritime Discoveries – Opportunities for a New Ocean Geopolicy," *Geo-Journal* 26:4 (1992): 477–81.

Vitet, Ludovic. *Études sur l'histoire de l'art* (Paris, 1868).
Monographie de l'église Notre-Dame de Noyon (Paris, 1845).
Histoire des anciennes villes de France: Dieppe (Paris, 1833).

Voigt, Lisa. *Writing Captivity in the Early Modern Atlantic: Circulations of Knowledge and Authority in the Iberian and English Imperial Worlds* (Chapel Hill, 2009).

Walker, Timothy. "Acquisition and Circulation of Medical Knowledge within the early Modern Portuguese Colonial Empire," in Daniela Bleichmar, Paula De Vos, Kristin Huffine, Kevin Sheehan (eds.), *Science in the Spanish and Portuguese Empires, 1500–1800* (Stanford, 2008).

Wallis, Helen (ed.). *The Maps and Text of the Boke of Idrography, presented by Jean Rotz to Henry VIII; now in the British Library* (Oxford, 1981).

Waters, David W. *The Sea or Mariner's Astrolabe* (Coimbra, 1966).

Weiss, Nathanaël. "Note sommaire sur les débuts de la Réforme en Normandie (1523–1547)," *Congrès du millénaire normand* 1 (1911): 193–205.

West, Ashley. "Global Encounters: Conventions and Invention in Hans Burgkmair's Images of Africa, India, and the New World," in Jaynie Anderson (ed.), *Crossing Cultures: Conflict, Migration, Convergence.* Proceedings of the 32nd Congress of the International Committee of the History of Art (Melbourne, 2009), 272–279.

Westman, Robert. "Proof, Poetics, and Patronage: Copernicus's Preface to *De revolutionibus*," in D. Lindberg and R. Westman (eds.), *Reappraisals of the Scientific Revolution* (Cambridge, UK, 1990), 167–205.

"The Astronomer's Role in the Sixteenth Century: A Preliminary Study," *History of Science* 18 (1980): 105–147.

Whitt, Laurelyn. *Science, Colonialism, and Indigenous Peoples: The cultural politics of Law and Knowledge* (Cambridge, 2009).

Wiedemann, Thomas. "Sallust's 'Jugurtha': Concord, Discord, and the Digressions," *Greece & Rome* 40:1 (1993): 48–57.

Williams, Elizabeth A. "Art and Artifact at the Trocadero," in George Stocking, Jr., *Objects and Others: Essays on Museums and Material Culture* (Madison, 1988).

Wintroub, Michael. "Translations: Words, Things, Going-Native and Staying True," *American Historical Review* 120:4 (October 2015): 1185–1217.

"Taking a Bow in the Theater of Things," *Isis* 101:4 (December 2010): 779–793.

A Savage Mirror: Power, Identity and Knowledge in Early Modern France (Stanford, 2006).

"L'ordre du rituel et l'ordre des choses: l'entrée royale d'Henri II à Rouen (1550)," *Annales: Histoire, Sciences Sociales* 56 (2001): 479–505.

"Taking Stock at the End of the World: Rites of Distinction and Practices of Collecting in Early Modern Europe," *Studies in History and Philosophy of Science* 30:3 (1999): 395–424.

"The Looking Glass of Facts: Collecting, Rhetoric and Citing the Self in the Experimental Natural Philosophy of Robert Boyle," *History of Science* 35 (June 1997): 189–217.

Wiseman, Timothy. *New Men in the Roman Senate, 139 B.C.–14 A.D.* (Oxford, 1971).

Young, Sandra. "Envisioning the Peoples of 'New' Worlds: Early Modern Woodcut Images and the Inscription of Human Difference," *English Studies in Africa* 57:1 (2014): 33–54.

Yusof, Yusharina, Zulkifli Ab Ghani Hilmi, and Saharani Abdul Rashid. "Weight Measurements in the Malay Minangkabau Culture," 2010 International Conference on Science and Social Research (CSSR 2010), December 5–7, 2010, Kuala Lumpur, Malaysia, 861–866.

Zanier, Giancarlo. "Platonic Trends in Renaissance Medicine," *Journal of the History of Ideas* 48:3 (1987): 509–519.

Zeller, Gaston. "Une notion de caractère historico-sociale: La dérogeance," *Cahiers internationaux de sociologie*, n.s. 22 (1957): 40–74.

Zetterberg, J. Peter. "The Mistaking of 'the Mathematics' for Magic," *Sixteenth Century Journal* 11 (1980).

Zilsel, Edgar. *The Social Origins of Modern Science*, ed., D. Raven, W. Krohn, and R. Cohen (Dordecht, 2003).

Zumthor, Paul. *Le masque et la lumière: la poètique des rhètoriqueurs* (Paris, 1978).

Zupk, Ronald Edward. *Revolution in Measurement: Western European Weights and Measures Since the Age of Science*, Memoires of the American Philosophical Society, Vol. 186 (Philadephia, 1990), 3–24.

Index